SMALL-SCALE FOOD PROCESSING

SMALL-SCALE FOOD PROCESSING

A DIRECTORY OF EQUIPMENT AND METHODS

SECOND EDITION

Compiled by

Sue Azam-Ali
Emma Judge
Peter Fellows
Mike Battcock

Practical Action Publishing Ltd
27a Albert Street, Rugby, CV21 2SG, Warwickshire, UK
www.practicalactionpublishing.org

First published in 1992
Second edition published in 2003

ISBN 978 1 85339 504 8

Technical Centre for Agricultural and Rural Cooperation (ACP-EU)
The Technical Centre for Agricultural and Rural Cooperation (CTA) was established in 1983 under the Lomé Convention between the ACP (African, Caribbean and Pacific) Group of States and the European Union Member States. Since 2000 it has operated within the framework of the ACP-EC Cotonou Agreement.

CTA's tasks are to develop and provide services that improve access to information for agricultural and rural development, and to strengthen the capacity of ACP countries to produce, acquire, exchange and utilise information in this area. CTA's programmes are organised around four principal themes: developing information management and partnership strategies needed for policy formulation and implementation; promoting contact and exchange of experience; providing ACP partners with information on demand; and strengthening their information and communication capacities.
CTA, Postbus 380, 6700 AJ Wageningen, The Netherlands

ITDG acknowledges the support of

A catalogue record for this book is available from the British Library.

Since 1974, Practical Action Publishing has published and disseminated books and information in support of international development work throughout the world. Practical Action Publishing is a trading name of Practical Action Publishing Ltd (Company Reg. No. 1159018), the wholly owned publishing company of Practical Action. Practical Action Publishing trades only in support of its parent charity objectives and any profits are covenanted back to Practical Action (Charity Reg. No. 247257, Group VAT Registration No. 880 9924 76).

Typeset by Dorwyn Ltd

Contents

Acknowledgements

This guide could not have been compiled without the efforts of many people. Thanks are due to the many company managers and directors around the world who took trouble to reply to our letters and without whose information, photographs and diagrams there would be nothing to publish; also to the companies that responded but whose equipment was outside the scope of the guide. We would particularly like to thank the authors of the first edition – Peter Fellows and Anne Hampton – for the original research into equipment providers and manufacturers, which made this aspect of the work easier the second time round.

Thanks are also due to Barrie Axtell for researching new suppliers and manufacturers, to Matt Whitton for preparing the artwork, to our colleagues at the Groupe de Recherche et d'Echanges Technologiques (GRET) and the FAO Information Network on Post-harvest Operations (INPhO) for sharing their technical information, and to other staff in ITDG for their support and assistance throughout.

Finally we are particularly grateful to Northern Foods and the National Lottery Charities Board (NLCB) which provided financial support for the preparation and publication of the guide.

Preface

The level of interest in small-scale food processing in developing countries has increased dramatically in recent years. This is partly because of greater food insecurity, particularly in Africa, with the need to preserve foods against drought, and partly as a result of the promotion of food processing to help increase incomes and employment. In other regions the success of agricultural development programmes in some countries has produced food surpluses that require preservation and processing. As a result there has been a corresponding upsurge in enquiries about the availability of low cost, small-scale equipment and where it can be found.

To respond to this need, ITDG produced the first edition of this guide in 1992. Feedback from users has indicated the value of the information to development workers and food processing entrepreneurs alike, and ITDG has now completely revised and updated the first edition.

In food processing it is necessary to select equipment that has a similar throughput to other machines used in different stages of a process that precede and follow it. If one piece of equipment is too small it will cause delays in the process and food may spoil. If equipment is too large, it is a waste of money. It is therefore necessary to look at the *whole* process when deciding on the equipment required, and for this reason we have included chapters that describe the stages and equipment needed to process selected foods in the first part of this book. The second part catalogues the different sizes and types of equipment that are available.

The selection of equipment was not easy and we make no claim to be comprehensive. The overriding principle is that equipment should be suitable for micro-, small- or medium-scale processors in developing countries. So you will not find the complex, automatic, continuous equipment used by high-technology food processors in industrialized countries.

However, the scale of production, the amount of money available to invest, and the availability of services (electricity, clean water, gas, maintenance facilities etc.) obviously vary in different countries and even between regions of the same country. We have therefore included wherever possible a range of equipment from simple hand-operated tools through to larger factory-produced machines. One of our aims has been to include equipment manufactured in developing countries to promote direct South-to-South trade relations. It should be noted that the response to our request for information was overwhelming from manufacturers in India, but elsewhere it was modest, particularly from African countries. We would be grateful to hear from manufacturers who are not included in the guide in order to improve our database for future editions.

We hope that this expanded and updated second edition of *Small-Scale Food Processing* will form a useful addition to the relatively limited information available on food processing equipment and serve its purpose as a tool to assist the improvement in peoples' livelihoods throughout the developing world.

Introduction

The importance of small-scale food processing

Throughout history, foods have been processed to improve their taste and appearance, to make them safe for consumption, or to preserve them for the off season. The basic techniques of food preparation and processing are widely known and are used every day in the preparation of family meals. However, food processing as a scientific and technological activity covers more than just food preparation and cooking. It involves the application of scientific principles to slow down or stop the natural process of food decay caused by micro-organisms or enzymes in the food. It also uses science together with the creative imagination of the processor to change the taste, texture and appearance of foods to provide people with interesting new products that add variety to their diet.

Traditional food processing techniques have been passed on through families for generations. Some have been modified and become more sophisticated as the equipment and processes are developed. Most regions have their own speciality foods that are based on local materials and tastes, and are unique to that region. This local distinctiveness is slowly being eroded in some countries, and more uniform types of foods are being produced across the world. A particular advantage of small-scale producers is their ability to produce traditional foods and develop niche markets that larger manufacturers are unable or unwilling to enter. Processing may also be used to create new foods and to add value to raw materials. This last point is particularly important where processing is used for employment and income generation.

The reasons for food processing are varied. In some countries the main aim is to preserve basic foods such as cereals and vegetables for use in the off season, thus contributing to food security. It can also be used to improve the nutritional quality and safety of foods. Others wish to process foods to create employment and to generate income, either as a little extra money to supplement household income or to establish and expand a recognized food processing business. And for others food processing is an attractive means of gaining income and improving their livelihoods. Small-scale food processing thus has the potential to bring many benefits to people in developing countries.

Improved nutrition and food security through household preservation

Improved home preservation for both household consumption out of the harvest season and for sale or exchange makes a significant contribution to household food security. Traditionally in areas that have regular dry seasons or droughts, processed foods are essential for survival in rural households (Campbell, 1987).

Simple, traditional methods of processing are used for the nutritional improvement of foods. For example, germination and toasting of cereal grains improves their digestibility and is used in the production of infant foods. Some food processes increase the bio-availability of nutrients by removing anti-nutritional compounds, such as the removal of cyanide compounds from cassava during the production of *gari*.

Processing also has the potential to add variety to the diet. In most societies the diet is based on a starchy staple (e.g. rice, cassava,

> **Examples of enhanced nutrition using processed foods**
>
> *Gundruk* is a dried fermented vegetable product, made in Nepal by shredding mustard, radish and cauliflower leaves and fermenting them in an earthenware pot for up to seven days. After drying in the sun they are stored for use over the winter period. The product is prepared as an appetizer and side dish with the main rice meal. *Gundruk* is crucial to remote communities for ensuring food security, and is a valuable source of vitamins and minerals when there are few fresh vegetables available (Karki, 1986).

maize, wheat or potato). By themselves these staples contribute calories and bulk to satisfy the hunger but offer little else. Diets are made much more appetizing and nutritious by the use of side dishes such as pickles, soups and sauces made from vegetables, fruits, fish or meat.

Income generation and employment

Food processing is probably the most important source of income and employment in Africa, Asia and Latin America. For instance in sub-Saharan Africa it is estimated that between one- and two-thirds of value-added manufacturing is based on agricultural raw materials, and in some countries more than 60 per cent of the workforce is involved in the small-scale food processing sector (World Bank, 1989; Conroy et al., 1995; Dietz, 1999). Small-scale food processors often start by working from home using domestic equipment; they often have little money to invest in equipment, and difficulty accessing credit or loans. However, they must be able to produce uniform quality foods under hygienic conditions.

Some people see food processing as their main source of income. They are entrepreneurs who will take out a loan to buy specialized equipment and secure working capital and, if successful, they will develop their business and marketing skills to expand and diversify their enterprise.

Formal businesses (i.e. registered with the authorities) and informal enterprises are often located in urban areas or peri-urban suburbs of large cities, where they provide employment for expanding urban populations as well as meeting the food needs of city dwellers who do not have access to land. This type of employment is becoming

Food processing is an important source of income and employment

increasingly important in many countries where changes to economic policies force large-scale retrenchment from government service, or traditional manufacturing sectors are unable to absorb the growing labour force. Small-scale processors also play an important role in generating income and employment in rural areas. This may be to supply part-processed foods to urban manufacturers or to sell foods locally in village markets. It is widely thought that rural populations have little disposable income for processed foods, but the large numbers of people that live in villages and rural towns mean that even if the individual purchases may be smaller than in urban areas, the total demand is very large.

> **Small-scale food processing is appropriate for women**
>
> In most societies, household preparation and processing of food is the responsibility of women. Compared to other types of manufacturing, the investment required to set up most food processing operations is small, and there are thus good opportunities for women to develop their household skills and establish successful businesses. Throughout the developing world there are numerous examples of women who have utilized their skills and established a processing business to support their families.

Whatever the reason for processing food, it is anticipated that this book will advise on the most appropriate equipment needed.

Scales of processing enterprises

The size of businesses described in this book are defined as follows:

- A **micro-scale** food processing enterprise is carried out by individuals for subsistence or sale in nearby markets. Typically, the number of workers is less than 10, the investment does not exceed $20 000 and annual sales are less than $12 000.
- A **small-scale** enterprise is one in which the owner may work on site or employ a manager. The number of workers is less than 20, the investment is less than $50 000 and annual sales are less than $25 000.
- A **medium-scale** enterprise is generally carried out by a group of people who pool their resources, or by a more wealthy individual who has sufficient money to invest in raw materials and equipment. Processors may grow raw materials themselves or contract farmers. Typically the number of workers is around 50, the investment is up to $500 000 and annual sales are more than $100 000.
- **Large-scale** food processing is characterized by a high degree of mechanization and low labour requirements. It requires a large capital investment, high technical and managerial skills and a substantial supply of raw material for economical operation. In developing countries large-scale processing is mostly confined to the brewing and sugar sectors and a few food processes that cannot be carried out on a small scale (for example, solvent extraction of cooking oil and concentration of fruit juices).

All scales of food processing have a place in developing countries but micro-, small- and medium-scale enterprises are more widespread than large-scale plants, and these are therefore the focus of this book.

Environmental considerations

In general, food processing is less environmentally damaging than many other forms of manufacturing and, again in general, the smaller the scale of processing the less environmental impact it has. Many micro- and small-scale processes use manual procedures and have low levels of energy consumption, the main exceptions being bakery (fuel for ovens) and milling (power for mills). Many small plants obtain their raw materials and distribute their products locally, and therefore do not incur significant fuel costs for transport. Similarly, most small-scale producers do not use sophisticated packaging materials that require energy-intensive production.

A local environment is not overloaded by the small volumes of waste produced by micro- and small-scale food processing, and it is able to detoxify them without causing pollution. Possible exceptions to this are starch processing and abattoirs/meat processing or dairies, where particularly polluting wastes may cause localized pollution if not properly handled and treated. Pollution from food packaging (particularly polythene bags) is however becoming an issue in many cities in developing countries. Further details of environmental considerations are given in a number of other publications (Fellows, 2000; Fellows and Axtell, 2002).

The selection of appropriate technology is an important consideration for small-scale processors

Selection of technology

The selection of suitable products, and the process by which to make them, requires very careful consideration. It is not sufficient to assume (as many people do) that a viable small business can be created to use up a surplus of raw material and prevent food wastage. There has to be a demand for the processed food that is clearly identified before a business is established.

Processors must therefore carry out a market survey and feasibility study to determine the viability of all potential food businesses.

To conduct a feasibility study a processor should:

- carry out market surveys to determine which processed foods are in demand and the price people will pay for them
- set the scale of production to meet a predetermined proportion of this demand
- decide the size and type of equipment required for this scale of production (also using technical advice on the best way to process food).

One of the purposes of this guide is to show what equipment is currently available and the approximate price of each piece. Further details of how to conduct feasibility studies are available in a number of other publications (Young and MacCormac, 1986; Kindervater, 1987; Fellows et al., 1996; Fellows, 1997a).

In general, the types of products that are most suitable for production at a small scale are those that have a high value added by processing. This means that for a particular level of income, the amount of food that must be processed is relatively small, and the size and type of equipment required can be kept at affordable levels. Typically, snack foods made from cereals and root crops, dried fruits, herbs and spices, or pickles made from vegetables, each use cheap raw materials, and can be simply processed into a range of goods with a considerably higher value.

A second important consideration when selecting a processing technology and product is to choose one which is inherently safe and has a low risk of causing food poisoning. Food is the only commodity that people buy every day and take into their bodies, and the overriding concern in all food processing is to avoid food poisoning. Acidic foods such as yoghurt, pickles, fruit juices and jams and most types of dried foods are considered to be safe, as they have a low risk of carrying food-poisoning organisms. In contrast, low acid foods such as meat, milk, fish and some vegetable products are much more susceptible to transmitting food-poisoning organisms through incorrect processing and poor handling conditions. Further details of food poisoning and high/low acid foods are given in Part I. It is essential for processors to either have experience of processing low acid products, or to have received training before they start production.

Some types of processing carry a high level of risk (for example, canning, bottling, chilling and freezing of low acid foods) and if they are not carried out correctly they can cause serious food poisoning. However, these processing methods are also fairly costly to establish and are therefore unsuitable for use by small- and medium-scale processors.

Special problems of food processing enterprises

Food processing has a number of special problems which make it different from most other types of business. For example:

- Most raw materials are highly perishable and spoil quickly after harvest or slaughter unless processed.
- They may also be highly seasonal which means that they can only be processed for part of the year, or alternatively they are part-processed for intermediate storage until they are needed.
- Foods are biological materials whose composition varies with different varieties, soil types, climate and weather, pests and diseases. This can mean unpredictable supplies and costs for raw materials.
- Some processed foods also have a seasonal demand (e.g. for festivals and ceremonies), which further complicates the business of food processing.
- Even after processing, foods do not keep indefinitely. The shelf life of processed foods can vary from a few days to several months or years. The distribution and sales methods used by the processor must be suited to the expected shelf life of the food and be carefully organized so that customers receive the food before it spoils.
- Packaging is an important means of controlling the shelf life of foods, but there are universal problems in finding suitable packaging materials in developing countries. Technically advanced plastic films, cartons etc. usually have to be imported and can be very expensive.
- Processors and processing methods must meet strict standards of cleanliness and quality assurance to avoid the risk of harming or even killing customers by allowing the growth of food-poisoning organisms in products.

In no other type of business do processors operate under these multiple complex technical constraints. However, food processors do share with other types of small business the difficulties of operating in a frequently hostile economic environment. They are rarely formed into associations, and have little economic power or ability to seek the government support that is available to larger companies (for example, subsidies, foreign exchange allowances, price stabilization). They often need intermediaries, such as extension agents, to guide them to appropriate solutions for their own individual problems.

Support to small-scale processors

Despite these problems, many governments and development agencies promote food processing as a means of generating employment and alleviating poverty. The reasons are not hard to find: food is familiar to the target groups, the raw materials are readily available (often in surplus), technologies are accessible and affordable for small-scale operations, equipment can often be manufactured locally, thus creating further employment, and if the products are chosen correctly they have a widespread demand. It is the concern of support agencies to ensure that any changes that come from promotion of food processing benefit disadvantaged groups rather than further threatening their livelihoods. We are aware that the staff of support agencies are most likely to read this book and not the ultimate beneficiaries, small-scale food processors. These staff therefore have a responsibility to carefully evaluate the technologies described in this book to ensure that they meet the needs of individual processors who are being assisted. Another function of this book is to prompt ideas for local adaptations of equipment to meet particular needs. If a processor has a problem finding suitable equipment, it is hoped that the book can give ideas from which a local workshop owner can fabricate suitable equipment.

The criteria that will help in deciding whether to recommend a technology are complex and interrelated but are likely to include the following:

- technical effectiveness (whether equipment will do the job required at the indicated scale of production)
- flexibility to perform more than one function
- costs for both purchase and maintenance of equipment, spare parts and any ancillary services required
- operating costs and overall financial profitability
- health and safety features
- conformity with existing production conditions and compatibility with other parts of a process
- social effects such as displacement of a workforce
- training and skill levels required for operation, maintenance and repairs
- environmental impact such as energy consumption or pollution of air or local waterways.

It is stressed that each of these factors will have a different weighting in different circumstances. There can be no simple solution to the difficult task of weighing up all factors in a particular situation and making the 'best fit' from the available technologies. The factors should therefore be treated as an aid to judgement by support staff and not simply a checklist.

If adopted the technologies and equipment described in this book are likely to affect the economic status of many people, and not always positively. However, in comparison with the large-scale, automated technologies used by food processors in industrialized countries, those presented here are relatively benign. The gain in employment and sparing use of resources make these technologies more sustainable and therefore ultimately more valuable to the national economies of developing countries.

How to use this book

The guide is intended for use by the following people:

- development workers who wish to purchase equipment in order to set up or expand a food processing unit
- advisors, government officials or development agency staff who wish to know the kind of equipment needed to process particular types of food
- those who wish to know the range of small-scale food processing equipment currently available.

The guide can be used in the following ways:

- to find out what equipment is needed to process a particular food, and the approximate cost
- to find the name and address of a manufacturer for a piece of equipment
- to learn more about the potential uses of various types of equipment.

The guide is arranged in four parts:

Part I describes the basic principles of food preservation and processing.

Part II describes the processing of different food groups.

Part III is a directory of equipment and manufacturers/suppliers of equipment.

Part IV is a reference section with glossary, references and index.

In Part I the individual chapters cover aspects of food spoilage, basic principles and methods of food preservation, food safety, hygiene and quality assurance issues, and the basic details of packaging.

In Part II separate chapters describe the processing of different commodity groups (e.g. fruit and vegetables, dairy products, meat and fish). Each chapter has a description of the preservation principles for the commodities, some observations on suitability for small-scale processing and a description of the process and the equipment required for various products. Reference numbers link items of equipment described to the entries in the directory.

In Part III there are two sections – a directory of equipment and one of manufacturers or suppliers of equipment. The equipment directory is arranged in 43 subsections, each of which describes the main features of equipment and the range available. Some of the equipment in the directory is illustrated using line drawings.

It is important to recognize that illustrated examples have been selected as representative of the equipment being described, and this does not imply superiority in any way over similar equipment from other manufacturers.

The names and countries of manufacturers are listed below each group of equipment.

The manufacturers/suppliers directory is arranged by region and country. Suppliers' names and addresses are listed in alphabetical order within the country.

If you wish to know the type of equipment needed to produce a particular food:

1. look up the food on the contents page or in the index
2. turn to the appropriate page in Part II to see the list of equipment in the middle column of the relevant table. For example:

Processing stage	*Equipment*	*Section reference*
Prepare raw material	Fruit and vegetable cleaners	7.1
	Peeling machinery	29.0
	De-stoners	13.0

If you want to know where to buy the equipment and its approximate cost:

1. make a note of the equipment directory section numbers in the right-hand column of the table
2. look up each section in turn in the directory to find the names and countries of suppliers and the price code for the equipment. For example:

29.0 Peeling equipment

Manually operated and powered peeling equipment is available.

POTATO PEELER
Suitable for peeling any size potatoes for further processing into chips etc. Suitable for small-scale processing units.
Power source: manual
Price code: 1
Central Institute of Agricultural Engineering, India

Note: in many cases a range of equipment will be shown and it is necessary for you to select the most appropriate type and size for your needs. All measurements (dimension and capacity) are quoted as supplied by the manufacturers. You should check direct, where necessary, for metric or imperial equivalents.

The price codes used throughout this guide are as follows:

Price code	Cost ($US)
1	0–200
2	201–1000
3	1001–5000
4	>5001

To find out which manufacturers supply equipment in a particular country, look up the manufacturers' directory and find the country (in alphabetical order within region). Manufacturers' names and addresses are given under each country (also in alphabetical order).

When you have found the equipment you need, we suggest that you write directly to the manufacturer for an up-to-date specification, current FOB (free on board) price for delivery to your country, delivery times and cost of spare parts. Remember, it will help the manufacturer to reply quickly if you include the full name of the equipment and model number where available. The more information that you can give about your needs, product throughput, voltage and so on, the better the manufacturer can provide you with the right tools for the job. Please note that ITDG is not able to supply any equipment.

We have made every attempt to ensure that the information in this guide is accurate but inevitably changes will have taken place since compilation. We ask you to contact us with any corrected information you are given so that we may update our records for future editions.

Please write to:
Agro Processing Specialist
ITDG
Schumacher Centre for Technology and Development
Bourton on Dunsmore
Rugby CV23 9QZ
UK
Fax: +44 (0)1926 634401
E-mail: itdg@itdg.org.uk

This guide relies on information supplied by manufacturers, and inclusion of an item of equipment is not a recommendation by ITDG, or a guarantee of its suitability or performance.

While every effort is made to ensure the accuracy of the data, the publishers and compilers cannot accept responsibility for any errors which may have occurred.

Please note that equipment availability and specifications are subject to change without notice and these should be confirmed with suppliers when making enquiries or placing orders.

PRINCIPLES OF FOOD PRESERVATION

Food spoilage

From the moment foods are harvested or slaughtered they deteriorate and eventually become unfit or unsafe for consumption. One of the main purposes of food processing is to prevent or slow down this deterioration. Food spoilage is brought about by a combination of physical damage (e.g. bruising or cuts to the surface of fruits and vegetables), chemical and biochemical reactions (e.g. development of rancidity in fats or colour changes in fruits and vegetables due to enzyme activity), and changes caused by micro-organisms, such as the growth of mould or slime on the food surface and changes in texture and taste of foods.

Physical damage

Raw foods, especially fruit and vegetables, are very susceptible to physical damage. Poor handling at harvest or during transport leads to damage to the tissues, which are then more susceptible to biochemical changes and invasion and spoilage by micro-organisms. Physical damage can be reduced by careful post-harvest handling and storage, and the use of containers instead of dumping foods in piles or in the backs of vehicles.

Enzyme activity

Enzymes are present in all living tissues and control the growth and development of the plant or animal. They continue to act after harvest or slaughter, causing undesirable changes in the appearance, taste and texture of foods. The ripening of fruits after harvest is also due to enzyme activity. When most plant materials are cut and exposed to air, enzymes present in the cells cause colourless chemicals to be converted into brown coloured compounds (known as enzymic browning). Enzymes that are produced by micro-organisms cause similar biochemical changes to foods to produce unpleasant and sometimes toxic products.

Damage caused by enzyme activity can be controlled by heating the food to denature the enzymes. Boiling, frying, pasteurization and canning are all heating methods that reduce enzyme activity. Blanching of vegetables is used before other forms of processing such as drying or freezing which do not heat the food sufficiently to destroy the enzymes. Alternatively, enzyme activity can be inhibited by changing the level of acidity, excluding air, or reducing the moisture content in some foods.

Chemical changes

Non-enzymic browning is a chemical reaction that takes place in foods, leading to browning of the material. In some instances, such as in the browning of bread crust and toasting of cereals, this produces desirable flavours and colours. However, in other foods such as dried milk it leads to the production of undesirable brown colours. It can be reduced by a number of methods: lowering the temperature of storage, optimizing the moisture content (the reaction rate is lowest at very high and very low moisture contents), increasing the acidity or using chemicals such as sulphur dioxide.

When foods are exposed to air, fats and oils are prone to oxidation which results in the development of off flavours and a reduction in nutritional value. This is a particular problem in oily fish and cooking oils. Oxidation is promoted by sunlight and some metals (e.g. copper and brass) which should not be used as containers for fatty foods. Oxidative changes can be prevented by excluding air with airtight packaging and keeping foods cool and away from light. At a more sophisticated level of technology, flushing foods with nitrogen or carbon dioxide gases before packaging in an airtight light-proof container can prevent oxidation.

Microbial spoilage and food poisoning

Micro-organisms live in abundance all around us in the soil, in water and air, and in the digestive tracts of animals. The three main types that are important in food processing are yeasts, moulds and bacteria, although viruses may also be important in particular foods (e.g. milk). Given the correct conditions for growth, micro-organisms multiply rapidly, causing undesirable changes to foods including changes to the taste, texture and appearance. Sometimes there is visible damage such as slime formation or the growth of moulds; often there is an offensive odour, particularly if proteins are broken down, and some foods may develop a bitter or acidic off flavour.

Microbial contamination of foods is wasteful but it can also be harmful, and in some cases life-threatening if food-poisoning bacteria (or 'pathogens') are present. They can be transmitted to foods by a number of routes: by animals, insects or birds that come into contact with foods, packaging materials or equipment; by poor personal hygiene of food handlers; or by cross-contamination from raw materials to a processed food via surfaces, utensils, equipment or human hands. Methods to control and prevent food poisoning by good hygiene and quality assurance are described in the chapter on *Food safety, hygiene and quality assurance*.

Food poisoning can be caused by eating the bacteria themselves, or by poisons (or 'toxins') that the micro-organisms release into the food. If this occurs and the micro-organisms are then killed by processing, the poisons can remain in the food unseen (for example, aflatoxin, which is produced by a mould in poorly dried cereals and nuts – see Part II). Some types of micro-organisms also produce dormant spores which can regrow when conditions become favourable. This can be a source of contamination and potential food poisoning in dried foods which are subsequently rehydrated. The symptoms of an attack of food poisoning can include stomach pains, diarrhoea, vomiting, headache, fever and aching limbs. Sometimes the illness lasts for days, weeks or months, and in some cases it can kill.

However, not all microbial activity in foods is undesirable. Yeasts are used to leaven bread and produce alcohol by fermentation of fruits and grains. A number of foods are preserved by the action of micro-organisms, especially lactic acid bacteria which are used in vegetable, meat and dairy fermentations (see Part II). Moulds are used for the production of *tempeh* from soybeans, in some types of cheese and for development of flavours in a number of other fermented foods.

The factors that control microbial growth, and hence the risk of spoilage or food poisoning, are temperature, acidity, moisture content, presence of air and the concentration of salts or sugar. Control of these factors is the basis of most methods of food preservation.

Temperature

Micro-organisms have an optimum temperature range (20–30 °C) in which they can grow, which corresponds to daytime temperatures in many developing countries. Outside of this range either they die or growth is inhibited. Therefore heating and cooling are used as methods of controlling microbial activity. For example, freezing to minus 18 °C prevents microbial growth (but does not necessarily destroy the cells); chilling to below 5 °C or holding cooked foods above 65 °C prevents most microbial spoilage and food poisoning. Heating foods to above 90 °C for several minutes is the basis for destroying micro-organisms by blanching and pasteurization. Heat sterilization (canning or bottling) and concentration by boiling (e.g. jams and sugar confectionery) use higher temperatures (e.g. 105–120 °C) and/or longer heating times to destroy nearly all micro-organisms in a food.

Acidity

pH is a measure of the strength of acid or alkali using a scale from 1 to 14, where 1 is very strong acid, 7 is neutral and 14 is very strong alkali. The majority of foods have a pH of 7 or less and are classified into three groups:

- low acid foods: pH greater than 5.3 (e.g. meats, fish, milk, root crops, vegetables)
- medium acid foods: pH 4.5 to 5.3 (some fruits, e.g. banana, pumpkin, papaya, melon)
- acid foods: pH below 4.5 (e.g. pineapple, citrus fruits, tomatoes)

Foods that have low acidity (a high pH) are more susceptible to bacterial spoilage and food poisoning. Beneficial lactic acid bacteria are tolerant of more acidic conditions (pH as low as 3.8), and yeasts and moulds are the most tolerant to acidic conditions, being able to grow at a pH as low as 2.5. The pH of foods can be adjusted by the addition of acids such as citric or acetic acid to prevent the growth of food-poisoning bacteria (e.g. in pickles and yoghurt).

Moisture content

Water is essential for the growth of all animal, plant and microbial cells, and if it is removed, or made unavailable, cellular activity is decreased. For example, removal of water by drying or changing it to ice during freezing makes it unavailable to microbial cells and hence preserves the food. High salt or sugar concentrations have a similar effect.

Bacteria require more water than yeasts, which in turn require more water than moulds for cell growth. The amount of water in a food that is available to micro-organisms is referred to as the 'water activity' (a_w). Pure water has an a_w of 1.0, most bacteria are inhibited below a_w of 0.9, most yeasts are inhibited below a_w of 0.8 and most fungi below a_w of 0.7. Almost all microbial activity is inhibited below a_w 0.6. In practice, this means that foods are dried, or the concentration of salt or sugar is increased, to a point where the available moisture cannot support microbial growth. Table 1 illustrates the range of a_w values and gives examples of the types of food in which they are found.

Table 1. Effects of different levels of water activity (adapted from Fellows, 2000)

a_w	*Phenomenon*	*Examples*
1.00		Highly perishable foods
0.95	Some bacteria and yeasts inhibited	Foods with 40% sugar or 7% salt, cooked sausages, bread
0.90	Lower limit for bacterial growth; some yeasts and fungi inhibited	Foods with 55% sugar, 12% salt; cured ham, medium age cheese
0.85	Many yeasts inhibited	Foods with 65% sugar, 15% salt; salami, mature cheese
0.80	Lower limit for most enzyme activity and growth of most fungi	Flour, rice (15–17% water), fruit syrups, sweetened condensed milk
0.75	Lower limit for salt-tolerant bacteria	Fruit jams
0.70	Lower limit for growth of most fungi that can tolerate dry conditions	
0.65	Maximum rate of non-enzymic browning	Rolled oats (10% water), molasses, nuts
0.60	Lower limit for growth of yeasts and fungi that can tolerate high sugar concentrations and dry conditions	Dried fruits (15–20% water), toffees, honey
0.55	Lower limit for life to continue	
0.50		Dried foods (a_w=0–0.55), spices, noodles
0.40	Maximum rate of oxidation	Whole egg powder (5% water)
0.25	Maximum heat resistance of bacterial spores	Crackers (3–5% water)
0.20		Whole milk powder (2–3% water), dried vegetables (5% water)

Methods of food preservation

Traditionally, the most widespread methods of food preservation have been dehydration, fermentation, salting, smoking and boiling. Other processes are controlled heating (pasteurization, canning etc.) and chilling and freezing. Table 2 describes the technologies that are commonly used worldwide. More recent advances, such as irradiation, ultra-high temperature (UHT) processing and extrusion, are generally not suitable for small-scale operations in developing countries and are not included in this book. Further information on these processes is available in food technology textbooks (Fellows, 2000).

Preservation by drying

Drying is used to remove water from foods to inhibit the growth of micro-organisms and to reduce the weight and bulk of food for cheaper transport and storage. When carried out correctly, the nutritional quality, colour, flavour and texture of rehydrated foods are slightly lower than fresh food. The colour of many fruits can be preserved by dipping them in a solution of 0.2–0.5 per cent sodium metabisulphite or by exposing them to burning sulphur in a sulphuring cabinet. However, these chemicals are not permitted in the United States and in some European countries, and if foods are intended for export, specialist advice should be sought. Blanching vegetables prior to drying preserves the colour and nutritional value by preventing enzyme activity.

At its most simple, drying foods can be carried out with a minimum of equipment using the heat from the sun. Although this is widely used to dry crops in the field, the lack of control and risk of contamination make it less suitable for food processing. More advanced dryers having various capacities and levels of complexity are available (Directory section 14.0) which have several advantages over sun-drying:

- the drying process is speeded up
- drying can be carried out in adverse weather conditions (rain, high humidity, no wind) and at night
- the temperature and rate of drying can be controlled, thus products of a consistent and higher quality are easier to produce.

However, if drying is carried out incorrectly there is a greater loss of nutritional and eating qualities, and more seriously a risk of microbial spoilage and possibly even food poisoning.

Principles of drying

Foods are dried when the water contained within them is removed into the surrounding air. It must first move to the surface of the food and then be evaporated as water vapour. For effective drying, the air should be *hot*, *dry* and *moving*. These factors are interrelated and it is important that each factor is correct (for example, cold moving air or hot, wet moving air are unsatisfactory).

The dryness of air is termed 'relative humidity' (RH), and the lower the humidity, the drier the air. Air with 0 per cent RH is completely dry air, whereas air at 100 per cent RH is fully saturated with water vapour. Air can only remove water from foods if it has the capacity to hold extra water vapour. If high RH (or wet) air is used, it quickly becomes saturated and cannot pick up further water vapour from the food. Humidity is affected by the temperature of the air. At higher temperatures the humidity is reduced and air can carry more water vapour. Normally the air in a dryer should be

Solar drying is a popular choice for home preservation

Table 2. Summary of different types of food processing

Principle of preservation	*Process*	*Typical products (section)*
Reduction in moisture content	Drying	Dried spices, nuts, fruits, vegetables, meats and fish
	Addition of sugar	Jams, jellies, osmotically dried fruits
	Addition of salt	Fermented vegetables, salted fish or meat
Increased acidity	Lactic acid fermentations	Pickled vegetables, yoghurt, cheeses, sourdoughs, meats
	Acetic acid fermentation	Vinegar production
	Addition of acid	Some types of pickled vegetables
Increased temperature	Blanching	Vegetables, as a pre-treatment before drying or freezing
	Pasteurization	Liquid foods – e.g. juices, milk, beer, soft drinks
	Boiling	Most foods, limited shelf life
	Canning	Most foods, not appropriate for small-scale processing
	Frying	Fried meat, fish, cereal products
	Baking/roasting	Baked goods, roasted meats
Reduction in temperature	Chilling, freezing	Most products, but not appropriate for small-scale processing if a chilled or frozen distribution chain is not available

10–15 °C above room temperature in solar dryers and at 60–70 °C in artificial dryers. The RH of air entering the dryer should ideally be below about 60 per cent. Further details of how to calculate air temperature and humidity are given in a number of textbooks (Fellows, 2000; Axtell, 2002).

When a food is to be dried it is necessary to carry out experiments to find the rate of drying. The information is used to find the time that food should stay in the dryer so that the moisture content is low enough to prevent spoilage. The rate of drying also has an important effect on the quality of the dried foods and (in fuel-fired dryers) on the fuel consumption. To find the drying rate, the food is weighed, placed in the dryer and left for 5–10 minutes. It is then reweighed and replaced in the dryer. This is continued until the weight of the food does not change. Typically, a drying rate of 0.25 kg/hour would be expected for solar dryers depending on the design and climate, and 10–15 kg/hour for artificial dryers. If the drying rate is lower than this, the air temperature or speed is too low and/or the RH is too high.

The sample of food is left in an airtight container for a day and checked to ensure that no further moisture has moved from the inside to the surface (if it has it is likely to be soft or even mouldy). Case hardening is the formation of a hard skin on the surface of fruits or fish which reduces the rate of drying and may allow mould growth. It is caused by drying too quickly during the initial period and can be prevented by using cooler drying air. Experiments with air temperature and speed can be used to select the best conditions for each food.

The moisture content of the food can be found using equipment (see Directory section 38.3) or by grinding the dried food to small pieces, weighing it and heating at 100 °C in an oven for four hours, then reweighing. The moisture content is found as follows:

$$\text{Moisture content (\%)} = \frac{(\text{initial weight} - \text{final weight}) \times 100}{\text{initial weight}}$$

The final moisture content of the dried food shows whether it will be stable during storage. When a satisfactory product is produced, the same temperature and time of drying are then used routinely in production.

To ensure safe storage of dried products, the final moisture content should be less than 20 per cent for fruits and meat, less than 10 per cent for vegetables and 10–15 per cent for grains. The stability of a dried food during storage also depends on its ability to pick up moisture from the air. Different foods pick up moisture to different extents, but the risk is greater in regions of high humidity. For hygroscopic foods which readily pick up moisture it is necessary to package them in a moisture-proof material.

Examples of moisture contents and a_w values for selected foods and their packaging requirements are shown in Table 3.

Small-scale drying equipment

Solar dryers operate by raising the temperature of the air, which reduces the humidity and also causes the air to move through the dryer, increasing the rate of drying. Food is enclosed in the dryer and therefore protected from contamination by dust, insects, birds and animals. The higher drying rate also permits a higher throughput of food and hence a smaller drying area. The dryers are waterproof and therefore food does not have to be moved when it rains.

Dryers can be constructed from locally available materials and are relatively inexpensive. The designs vary from very simple direct dryers to more complex indirect designs which have separate collectors and drying chambers (Directory section 14.0). Food is either exposed directly to the sunlight (in direct systems) or placed in the shade and has heated air passed over it (in indirect systems). Direct systems are used for food such as raisins, grains and coffee where the colour change caused by the sun is acceptable, but most foods need indirect sys-

Table 3. Moisture content of some common foods and their packaging requirements (adapted from Fellows, 2000)

Food	*Moisture content (%)*	*Water activity (a_w)*	*Degree of protection required*
Fresh meat	70	0.985	Package to prevent moisture loss
Bread	40	0.96	
Marmalade	35	0.86	
Rice	15–17	0.80	Minimum protection or no packaging required
Wheat flour	14.5	0.72	
Raisins	27	0.60	
Nuts	18	0.65	
Toffee	8	0.60	Package to prevent moisture uptake
Boiled sweets	3.0	0.30	
Biscuits	5.0	0.20	
Potato crisps	1.5	0.08	
Spices	5–8	0.50	
Dried vegetables	5	0.20	
Breakfast cereal	5	0.20	

A range of small-scale mechanical dryers is available

tems to protect their colour. Other types of dryers use electric or wind-powered fans to increase the speed of the air but these add to the capital and operating costs.

The main disadvantage of solar dryers is that they lack the control of artificial dryers and it is difficult to routinely produce high quality products. The extra investment in small solar dryers does not produce corresponding benefits compared to protected sun-drying. However, large fan-assisted solar dryers have greater potential. Also, the use of solar energy to preheat artificial dryers is important to save energy. Large (400 kg capacity) fuel-assisted solar dryers that operate despite the weather and at night are claimed to dry pineapple and other fruits profitably.

Artificial (or 'mechanical') dryers (Directory sections 14.2 to 14.6) use fuel to increase the air temperature and reduce the RH, and fans to increase the air speed. They give close control over the drying conditions and hence produce high quality products. They operate independently of the weather and have low labour costs. However, they are more expensive to buy and operate than other types of dryer. In some applications, where consistent product quality is essential, it is necessary to use artificial dryers. The cabinet dryer is similar in design to the solar type, but is heated by burning fuel or electricity. To be economical this type of dryer should be relatively large (1–5 tonnes). They are successfully used for drying high value products such as herbs, speciality teas and vegetables (Axtell and Bush, 1991). There is a large range of other types of artificial dryers but their capital costs are too high for most small-scale processors.

Preservation by fermentation

Indigenous fermented foods have been consumed for thousands of years all over the world and are strongly linked to a region's culture and tradition. Fermentation is a low energy preservation process which increases the shelf life and decreases the need for refrigeration or other forms of preservation technology. It is therefore a highly appropriate technique for use in developing countries, especially in remote areas where access to sophisticated equipment and power is limited (Battcock and Azam-Ali, 1999).

Three types of micro-organism (bacteria, yeasts and moulds) are used in food fermentations. The most important bacteria are the *Lactobacillus* species, which have the ability to produce lactic acid from sugars in milk or vegetables. Other important bacteria, especially in the fermentation of fruits and vegetables, are the acetic acid producing *Acetobacter* species. The most beneficial yeasts are the *Saccharomyces* species, especially *S. cerevisiae*, which are used for leavening bread and the production of alcohol. Important moulds include *Aspergillus* species which impart characteristic flavours to foods such as cheeses.

Principles of fermentation

Nearly all food fermentations are caused by more than one micro-organism, either working together or in a sequence. An organism that initiates fermentation will grow until it exhausts the nutrients or until its by-products inhibit further growth and activity. During this initial growth period, other organisms develop which are ready to take over when the conditions become intolerable for the former ones. For example, in vinegar production yeast converts sugars to alcohol, which is the substrate required by the *Acetobacter* species to produce acetic acid. Similar sequences of lactic acid bacteria occur during the fermentation of pickles, and there is a mixed culture of two types of lactic acid bacteria in yoghurt fermentation (Part II – Dairy Products).

With any fermentation it is essential to ensure that only the desired micro-organisms start to multiply and grow on the food. This has the effect of suppressing other non-desirable pathogens or spoilage micro-organisms which cannot survive in either alcoholic or acidic environments. The six main factors that influence the growth of micro-organisms are described in this chapter. A fermentation can be controlled by manipulating the temperature, oxygen concentration, nutrient level and pH to produce the required product. For example, yeasts ferment sugars to alcohol and carbon dioxide in the absence of air, so by controlling the level of sugar and ensuring an airtight fermentation tank (Directory section 43.0) it is possible to maximize alcohol production. In contrast, excess oxygen is required by *Acetobacter* species to convert alcohol to acetic acid and special aerated fermenters (Directory section 43.0) are used in this case for the production of fruit vinegars. (This same reaction also occurs in wines in the presence of excess oxygen, producing an unpleasant, vinegary, off taste in the wine.)

Preservation using sugar or salt

The principle of this type of preservation is that the salt or sugar binds water that is contained within the food, making it unavailable for micro-organisms to use (i.e. reducing its water activity). Spoilage bacteria will not grow in sugar concentrations of 40–50 per cent, but some moulds and yeasts are tolerant of higher concentrations, and a *minimum* sugar content of 68.5 per cent is needed for jams and preserves. However, there is a group of yeasts and bacteria known as osmophilics that are tolerant of high sugar concentrations and it is safer to take the sugar content of jams and preserves to 70 per cent. It is important not to exceed 72.5 per cent as there is then the danger of crystallization of the sugar in the preserve. A refractometer is essential to ensure the correct sugar concentration. These osmophilic yeasts and bacteria are abundant in premises where jam is made routinely.

Sugar is added to fruit in the preparation of preserves, jams, jellies and sweets and some of the water is removed by boiling, thus increasing the concentration of sugar. Other sweeteners such as golden syrup or honey can be used, but they may alter the taste or colour of the preserve. Glucose syrup, which has little or no flavour and is less sweet than sugar (sucrose), can also be used as a preservative.

Salt concentrations of 10–12 per cent are adequate to control the growth of most spoilage bacteria. However, certain bacteria, notably *Leuconostoc* species and *Lactobacillus* species, are able to grow in higher concentrations of brine. Two methods are used in salting:

- dry salting, in which the food is packed in salt and left to allow the salt to penetrate the tissues
- brining or pickling, in which the food is immersed in a brine solution (3–10 per cent salt) over a period of time. Brine is also used in the processing of shrimp, olives, cheese, bacon and hams.

Preservation by smoking

At the simplest level, a food is hung over a smouldering wood-chip or log fire. The outer layers become coated with deposits of tars, phenols and aldehydes which have powerful antibacterial action. In more sophisticated processing, smoke generators or special smoking kilns (Directory section 33.0) are used in which the heat and smoke generation can be controlled. Heat also has a drying effect on the food, thus increasing the preservative effect. As a general principle, the longer the period of smoking, the longer is the shelf life of the product. There are two different methods of smoking:

- cold smoking, where the temperature (35–50 °C) is not high enough to cook the food
- hot-smoking, where the temperature is above 60 °C and is high enough to cook the food.

Hot-smoking is often the preferred method because it requires less control than cold processing, and the shelf life of the product is longer. The main disadvantage of hot-smoking is higher fuel consumption.

Preservation by heating

Heating to preserve foods without significant changes to their quality is done by blanching, which involves heating vegetables in either boiling water or steam for a few minutes (Directory sections 2.0, 27.0), or by pasteurization. There are two types of pasteurization, both of which involve heating foods below 100 °C for different lengths of time depending on the particular food: the first is bulk processing in boiling kettles (Directory section 27.0) before packaging, and the second is in-container pasteurization (Directory section 28.0). The relative advantages and limitations of each are described in Part II.

Frying and baking are heating methods that both aim to change the eating qualities of foods as well as increasing their shelf life. Frying can be done using shallow pans that produce a variegated pattern of browning on the surface of foods, or by deep fat fryers (Directory section 27.2) that produce more uniform colour. A wide range of bakery ovens

Smoking is a simple method of food preservation traditionally used for meat and fish

is available (Directory section 25.0), and details of baking conditions are given in Part II. The chemical changes that take place in foods during frying and baking are extremely complex and result in the characteristic flavours and textures of specific products. The characteristic golden brown colour is produced by non-enzymic browning (Part I).

Preservation by cooling

Both chilling and freezing slow down microbial and enzyme activity without causing significant changes to the quality of foods. They are successful if the lowered temperature can be maintained throughout the distribution chain from processor to customer, which is a major problem in most developing countries. In addition, both refrigerators and freezers (Directory sections 8.1, 8.2) use significant amounts of power, and electricity supplies should also be reliable to ensure that the temperature of foods does not rise during power failures. If the food temperature is allowed to rise, there is a substantial risk of spoilage and food poisoning, and for this reason great care is needed if these processes are being considered.

Preservation by the addition of chemical preservatives

In addition to salt, sugar, lactic and acetic acids, alcohols and components of smoke (above), chemical preservatives include salts of benzoic acid, a number of other organic acids, sodium metabisulphite and sodium or potassium nitrate (Table 4). Their use is regulated by maximum permitted levels, which vary between countries, and food processors should check the local regulations at a bureau of standards or equivalent organization.

Table 4. Commonly used chemical preservatives (from Ihekoronye and Ngoddy, 1985)

Compound	*Comments*	*Commonly used levels (%)*	*Used in*
Sulphites and sulphur dioxide	Sulphur dioxide gas and the sodium or potassium salts of sulphite, bisulphite or metabisulphite are the most commonly used forms. Sulphur dioxide is most effective as an antimicrobial agent in an acid environment. Sulphurous acid inhibits yeasts, moulds and bacteria. Sulphur dioxide is the most effective inhibitor of browning of foods. It is used with dried fruits to maintain a bright colour. Levels above 500 ppm give a noticeable taint.	0.005–0.2	Fruit juices, dried fruit and vegetables, semi-processed products
Nitrite and nitrate	Sodium or potassium nitrite and nitrate are used in curing mixtures for meat to prevent off flavours, to preserve the colour and to inhibit microbial growth. Nitrites have been shown to be involved in the formation of low levels of cancer-causing nitrosamines in cured meats, and their use as a curing agent is under debate.	0.01–0.02	Meats
Sorbic acid	Sorbic acid and sodium and potassium sorbate are widely used to inhibit the growth of moulds and yeasts in a range of products. The activity of sorbic acid increases as the pH decreases. Sorbic acid and its salts are practically tasteless and odourless in foods when used at levels <0.3%.	0.05–0.2	Cheese, baked goods, fruit juices, wine and pickles, jams, jellies
Propionic acid	Propionic acid and calcium and potassium propionate are effective against moulds and some bacteria.	0.1–0.3	Fruits, cheese, vegetables
Benzoic acid	Benzoic acid, in the form of sodium benzoate, is a widely used food preservative. It occurs naturally in cranberries, prunes, cinnamon and cloves. It is well suited for use in acid foods. Yeasts are more susceptible to benzoates than are bacteria and moulds. Benzoic acid is often used in combination with sorbic acid at levels from 0.05 to 0.1% by weight.	0.03–0.2	Vegetable pickles, fruit juices, carbonated beverages, jams, jellies
Citric acid	Citric acid is the main acid found naturally in citrus fruits. It is widely used in carbonated beverages and as an acidifier of foods. It is a less effective antimicrobial agent than other acids.	No limit	Fruit juices, jams, other sugar preserves

Food safety, hygiene and quality assurance

Successful food businesses are dependent upon good food safety and hygiene, and have a special responsibility not to injure their customers. The main ways in which a producer can harm consumers are by selling food that:

- contains poisonous materials
- contains bacteria or moulds or the poisons they produce
- contains glass or other contaminants that could cause harm if eaten.

Safe food can be produced by **careful attention to hygiene** and by **good quality assurance**.

Good hygiene means careful attention to the cleanliness of the processing room and equipment, and the personal hygiene of food handlers. This prevents the bacteria that are present on equipment and the hands and clothes of food handlers from contaminating the food.

Good quality assurance means establishing procedures to monitor and control every aspect of production that could result in harm to a consumer or a loss of product quality. In practice this means:

- careful selection of good quality raw materials
- use of the correct processing conditions, such as the temperature and time of heating
- preventing materials such as dirt, metal, cleaning chemicals, glass and stones from contaminating the food
- the use of suitable packaging materials to protect the food after processing
- control over conditions of storage and distribution to the consumer.

These factors will ensure that only wholesome food is produced without any contaminants. Details of the specific types of quality assurance procedures for different foods are given in Part II and appropriate equipment for quality assurance is included in the Directory (section 38.0).

Food safety and food hygiene

In most countries there are laws which are designed to protect customers against poisoning and injury. In particular, low acid foods such as meat products, fish, seafood and dairy products are covered by laws that are more strict than for other products, because of the higher risk of food poisoning. Harmful micro-organisms that must be guarded against include bacteria, such as *Salmonella* species (in poultry, meat and eggs), *E. coli* (in animal products), *Listeria* species and *Campylobacter* species (in poultry, meat, milk and dairy products), and viruses, such as Hepatitis A (in shellfish from polluted areas, fruits and vegetables or salads prepared in unhygienic conditions).

Readers are advised to contact their local bureau of standards, ministry of health or other relevant government department to obtain full details of the specific laws of their country. However, in the end it is the customer who is the most effective food inspector. If consumers become ill from eating a food, or if the quality varies each time they buy a pack, they will not buy food from that producer again. It is therefore in the producers' own interest to make safe wholesome foods of consistent quality because customers will return to buy again.

Personal hygiene

The main sources of food poisoning are microbes from workers' hands, from dirty tools and work surfaces or from raw materials.

Food processors should pay strict attention to personal hygiene and wear protective clothing

All persons handling food should pay strict attention to good hygiene practices. This includes wearing clean clothing (including aprons, hats to cover the hair, gloves and footwear). Any clothing that could contact the food should be thoroughly cleaned, every day if necessary.

All cuts or wounds should be covered by a waterproof dressing and kept clean, even if they are not on the hands. Workers should not handle foods if they have a stomach upset or a skin disease, or if they are looking after someone else with these illnesses. They should be found work that does not involve contact with the food until they recover. Managers of food processing units should ensure that no-one smokes, eats or chews anything while preparing food, spits, coughs or sneezes over food as this spreads bacteria that contaminate the food. They should ensure that anyone who touches food always washes their hands properly beforehand, using soap and clean water, and especially after every visit to the toilet. Particular care is needed when handling raw meat or poultry and any other food stuff, to avoid cross-contamination.

Cleanliness of equipment and production areas

All work surfaces and equipment must be washed with hot water and detergent before, during and after production. If surfaces, chopping boards or utensils (such as knives) are used for different foods they should be washed between each use. Cloths and sponges used to wipe down surfaces, and towels used for hand drying, should be washed and sterilized regularly by boiling in water for at least 10 minutes. Equipment and utensils should be stored where they can be kept clean when not in use, and brushes and cloths should be hung up after use, and never left on equipment or foods to dry. Cleaning equipment, chemicals and detergents should be stored in a separate cupboard from foods and processing equipment. All the equipment should be kept in a good condition and properly repaired, to prevent accidents as well as contamination. Only food-grade plastic, ceramic, stainless steel or enamelled metal are suitable materials for use in contact with foods. Copper, iron, brass or pewter should not be used. Galvanized iron or aluminium is only appropriate to use for some foods, and wooden utensils or surfaces are difficult to clean.

In production areas, managers should ensure that a regular cleaning programme is implemented to prevent dirt from gathering on window ledges or around table legs and other inaccessible places. All floors and drains should be washed down daily. Good lighting in the processing area not only helps to stop accidents, but also makes cleaning and inspection easier. Ingredients and raw materials should be covered and stored off the ground to protect them from contamination by insects, rodents and birds and the bacteria they carry. Flies, rats, mice and other vermin contaminate food with bacteria from their droppings or their bodies. They also feed on excrement of all kinds and transfer the infection to food either by vomiting on it or by walking over it with infected feet. The processing room should have screens on doors, windows and drains to prevent animals or insects from entering the room. All cupboard doors and lids of storage containers should fit tightly, and table legs can be placed in pots of water or kerosene to stop ants crawling up them.

During production, any spillages of food should be cleared straight away, and wastes should not be allowed to accumulate on floors, in drains or on work surfaces. They should be placed in covered bins and emptied at regular intervals, and should be disposed of properly away from the processing site.

Further information on food safety and food handling is given in the references by Fellows et al. (1998) and Shapton and Shapton (1993). A selection of books that contain further information on the safe handling and processing of foods is given at the end of the chapter.

Quality assurance

The control of all aspects of processing that could affect the safety or quality of a food is known as 'quality assurance' (QA). This is not just a series of tests that are done on a food, but a management system that predicts potential problems and takes steps to prevent them occurring before they have a chance to affect food quality or safety. The workers in a food production unit are an integral part of the QA plan and they should be properly trained to implement it.

The basis of a QA plan is to identify points in a process where a lack of control could affect the quality of a food. These are known as control points, and where food safety is a factor they are known as critical control points. For food safety the QA will involve a hazard analysis, critical control point (HACCP) plan. The details of how to devise and implement a HACCP plan are described in several publications (Dillon and Griffith, 1995; Mortimer and Wallace, 1998), and information on specific products is included in Part II. Typical equipment for QA is shown in Table 5 and described in the Directory (sections 38.0, 39.0).

Table 5. Typical equipment required for quality assurance of food products

QA equipment	*Use*	*Directory reference*
Calibrated scoops and measures	Measuring out ingredients	39.0
Hydrometer	Measurement of alcohol or salt concentration	38.2
Moisture meter	Measurement of moisture content	38.3
pH meter	Measurement of acidity	38.6
Refractometer	Measurement of sugar content	38.4
Thermometer	Measurement of temperature	38.5
Weighing scales	Measuring out ingredients, checking fill-weights	39.0

Simple methods of quality assurance

The Pearson square – common calculations simplified

Many small-scale food manufacturers find difficulty with the calculations involved in mixing two raw materials together to give a final product of known composition (fat content, alcohol content, sugar content etc.) The use of the Pearson square makes this type of calculation very simple. The Pearson square was developed many years ago and found particular application in the dairy industry for standardizing milk and blending milks to give a final product with a standard fat content.

Before looking at some examples of the use of this tool it is very important to understand that it can only be used in a two component system, for example blending two wines (to give a known alcohol level) or two milks (to give a standard fat content). If more than two components such as protein level and fat level are involved, then more complex 'mass balance' calculations are required.

Use of the Pearson square

Producing a 10 per cent butter fat cream

In this example homogenized milk with a fat content of 3.5 per cent is to be mixed with a 20 per cent fat cream to give a light cream containing 10 per cent fat. In what proportions should they be mixed?

First draw a rectangle and label the two horizontal lines with the names of the two products used, as shown here:

Now enter the composition of each ingredient as shown below, putting the required final product fat content in the centre of the box:

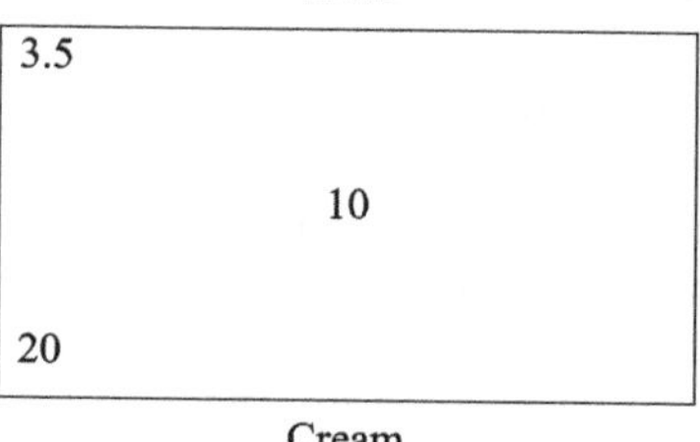

Mix the two components by crossing diagonally through the centre figure, and subtract each from the larger figure (3.5 from 10, and 10 from 20), giving 6.5 in the lower corner and 10 in the upper:

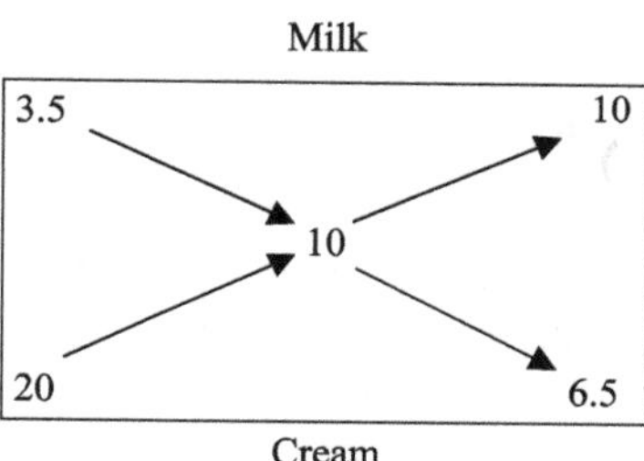

Read the results: these show that we need 10 parts milk (shown by the top line) to be mixed with 6.5 parts of cream (bottom line).

Production of pure sweetened fruit juice

In this example pure orange juice with a sugar content of 10 per cent is to be mixed with a 60 per cent sugar syrup to give a final sweetened juice containing 15 per cent sugar. Drawing the square as described above, it should look like this.

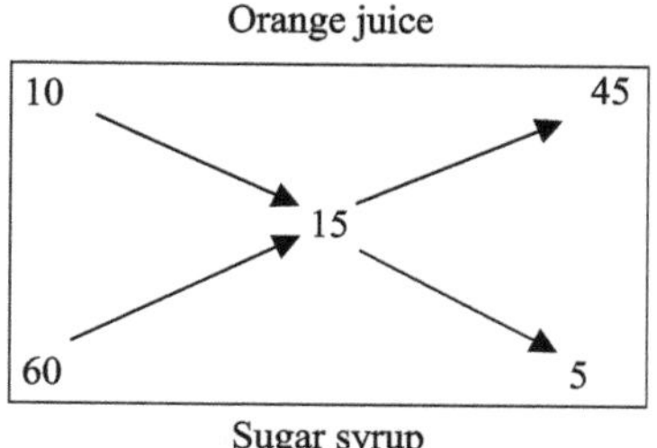

Subtracting diagonally shows that 45 parts (litres) of orange juice need to be mixed with

5 parts (litres) of sugar syrup to produce a final 15 per cent sugar sweetened juice.

Formulating a jam

In this example fruit pulp containing 10 per cent natural sugar is to be mixed with pure cane sugar to give a jam containing 70 per cent sugar.

Draw a square as before, enter the numbers and subtract diagonally.

The square should look like this:

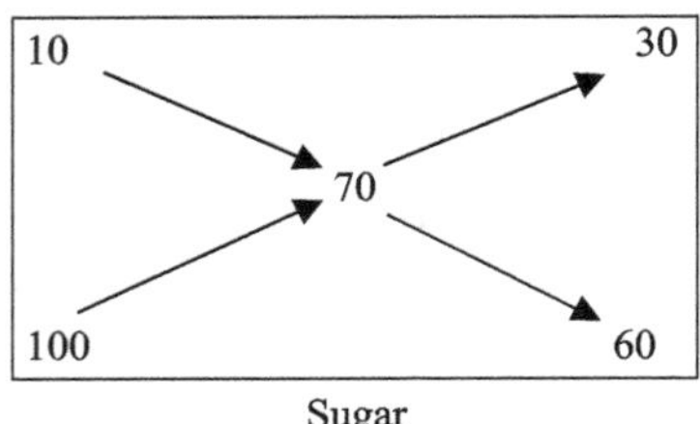

This indicates that 30 parts (kg) of fruit mixed with 60 parts (kg) of sugar will give a 70 per cent sugar jam.

Further reading

In the list of References see: Board (1988); Caplen (1982); Fellows et al. (1995); Hobbs and Roberts (1987); Howard (1979); IFST (1991); Kramer and Twigg (1962); Pickford et al. (1995); Shapton and Shapton (1994); Sprenger (1996); Steinkraus (1996); Water Research Centre (1989); Wood (1985).

In addition see the FAO Manuals of Food Quality Control (series published by the Food and Agriculture Organization, Rome, Italy):

Commodities (1979)
Food Analysis: General Techniques, Additives, Contaminants and Composition (1986)
Food Analysis: Quality, Adulteration and Tests for Identity (1986)
Food for Export (1979)
Food Inspection (1984)
Introduction to Food Sampling (1988)
Microbiological Analysis (1992)

Packaging

The main aims of packaging are three-fold:

- to contain the contents without leakage
- to keep the food in good condition until it is sold and consumed
- to encourage customers to purchase the product.

If adequately packaged, the shelf life of foods is extended, and this allows it to be distributed more widely, giving consumers more choice, and giving processors a larger potential market. Correct packaging also prevents wastage from leakage, or product deterioration during transportation and distribution.

Some foods, for example street foods and bakery goods, are expected to have a short shelf life and thus have minimal packaging requirements, and others, such as flour and sugar, are stable without sophisticated packaging. However, a wide variety of foods depend on packaging for their protection, and this is a major problem for food processors in most developing countries. There is a widespread lack of knowledge of available packaging materials and/or the requirements for packaging different foods. Each product has its own characteristics and packaging requirements vary according to the type of food, the climatic conditions under which it is stored and distributed, the expected shelf life and the requirements of the expected consumers. There is also comparatively little published information on food packaging for small-scale processors. The books by Fellows and Axtell (2002) and Obi-Boatang and Axtell (1995) are useful resource guides for small-scale packaging.

In some countries where a glass-making factory exists, bottles and jars are widely used, and in others there may be paper-processing plants that supply paper wrappers, cartons and cardboard boxes. For the majority of developing countries there are no facilities to make plastic films or containers and as a result these must be imported. Similarly, can-making plants are not found in most developing countries; cans are not widely used by small-scale processors because they are relatively expensive and cannot be reused.

In many parts of the world the shortage of packaging makes processors wrap foods in reused paper, leaves, rushes and other biodegradable wrappers. Although these materials are adequate for some products, consumer expectations are rising in all countries and these materials are not able to compete with more sophisticated materials used to pack imported foods or those from larger scale producers. There is thus no choice for most processors, except to pay higher prices for imported packaging materials. Packaging can thus represent a large part of the total cost of a processed food. This may be in part because of the higher unit cost when small quantities are ordered for small-scale production.

Functions of packaging

Packaging should provide the correct internal environment for a specific food from the time it is packed until its consumption. A suitable package should therefore perform the following functions:

- it should provide a barrier against dirt, micro-organisms and other contaminants, thus keeping the product clean
- it should protect a food against damage from crushing and the harmful effects of air, light, insects, and rodents
- it should be convenient for retailers and distributors to handle, and for the consumer to store and use

Packaging protects the product, while the label establishes the brand and identity

- it should help customers to identify the food and instruct them how to use it correctly
- it should help persuade the consumer to purchase the food.

Packaging materials

Common types of packaging material include:

- wrappers
- bags and sacks
- boxes
- bottles and jars
- pots
- plastic films
- metal cans.

Wrappers

Banana, vine or plantain leaves are used for wrapping certain types of food (e.g. steamed doughs and confectionery) and cooked street foods of all types. Corn husk is used to wrap corn paste or unrefined block sugar. Paper wrappers are used for some types of confectionery, salt, sugar and flour that are relatively stable (newsprint should not be used as the ink is toxic). These materials are a good solution for products that are consumed soon after purchase as they are cheap and readily available. However, they have poor marketing appeal and do not protect the food against moisture, oxygen, odours or micro-organisms, and are therefore not suitable for long-term storage. They readily absorb fats and moisture unless specially coated, and are fairly easily torn. For paper, some of these constraints can be overcome by treating it with wax, or impregnation with varnish or resin. Paper can also be strengthened by combining it with hessian cloth, cardboard or polythene. Wrapping machines are described in Directory section 26.3.

Bags and sacks

Vegetable fibres from bamboo, banana, coconut, cotton, jute, raffia, sisal and yucca are converted into yarn, string or cord and woven into bags and sacks. These materials are flexible, have strong resistance to tearing, have non-slip properties for stacking, and are lightweight for handling and transportation. Being of vegetable origin, all are biodegradable but also rot easily, so are only reusable to some extent. As with leaves, vegetable fibres do not provide protection to food that has a long shelf-life since they offer no protection against moisture pick-up, micro-organisms, or insects and rodents. Woven polypropylene sacks are steadily replacing fibre sacks in many countries as a result of their lower cost and resistance to rotting.

Boxes

Wooden shipping containers have traditionally been used for a wide range of solid and liquid foods, including fruits, vegetables, tea and beer. Wood offers good protection, good stacking characteristics, strength and rigidity. However, plastic containers have a lower cost and have largely replaced wood in many applications. The use of wood continues for some wines and spirits because the transfer of flavour compounds from the wooden barrels improves the quality of the product. Cardboard boxes are widely used for local distribution and storage. These can be either solid or corrugated construction.

Bottles and jars

Glass has many properties which make it a popular choice as a food packaging material:

- it is able to withstand heat treatments such as pasteurization and sterilization
- it does not react with food and is impervious to moisture, gases, odours and micro-organisms
- it is rigid and protects the food from crushing and bruising
- it is reusable, resealable and recyclable
- it is transparent, allowing products to be displayed; coloured glass may be used either to protect the food from light or to attract customers.

Despite its many advantages, glass does have certain constraints:

Glass jars and bottles can be recycled and reused

Bottle rinser

- it is heavier than many other packaging materials, leading to higher transport costs
- it is easy to fracture, scratch and break
- it is a potentially serious hazard due to glass splinters or fragments.

Most small-scale processors use reused glass containers. It is important that these are adequately cleaned as they are often used by consumers for storing a range of materials from kerosene to pesticides, and are therefore a potential hazard. Simple hand-held brushes can be used for cleaning at a small scale, and for higher production rates powered washers are available (see Directory section 3.0). Glass containers should be sterilized before filling by holding the open neck of the container over the spout of a kettle containing boiling water. Alternatively, containers can be inverted over a steam pipe which is supplied from a water boiler (Directory section 27.0). Caps and lids should not be reused as they do not form a proper seal. Suitable machines for sealing bottles and jars are described in Directory section 26.1. After heat processing, bottles and jars can be cooled using a simple cooler.

Plastic bottles and jars are increasingly available in many developing countries. Their advantages include lower costs, lower weight and resistance to breakage. Common materials include PET (polyethylene terephthalate), polypropylene and polyvinyl chloride (PVC). Sealing machines for plastic bottles and jars are described in Directory section 26.1.

Pots

Earthenware pots are used for storing foods such as curd, yoghurt, beer, dried food and honey. Corks, wooden lids, leaves, wax, plastic sheets or combinations of these are used to seal the pots. Unglazed earthenware is porous and is suitable for products that need cooling, such as curd. Glazed pots are needed for storing liquids (oils, wines) as they are light-proof and, if clean, restrict the entry and growth of micro-organisms, insects and rodents. However, earthenware is easily broken and does not have a good marketing appeal in many countries. In order to compete effectively many manufacturers use plastic pots, which are commonly made from polypropylene, high density polythene or polyvinyl chloride. They can be made at a small scale using a vacuum thermoforming machine and can be sealed with foil, plastic caps or films using small machines (Directory section 26.1). It is also possible to seal foil or plastic lids onto pots at a small scale using a domestic iron.

Plastic films

There is a very large range of flexible films that can be made with different thicknesses, gas and moisture resistance, light resistance and flexibility. Their advantages are:

- relatively low cost
- good barrier properties against moisture and gases
- they are heat-sealable to prevent leakage of contents
- wet and dry strength
- they are easy to handle and convenient for the manufacturer, retailer and consumer
- they add little weight to the product
- they fit closely to the shape of the product, thereby wasting little space during storage and distribution.

Commonly available imported films in developing countries include polythene, polypropylene and cellulose, but more complex laminated or co-extruded films are expensive and generally not available.

Low-density polythene is heat-sealable, inert, odour-free and shrinks when heated. It is a good moisture barrier, but has a relatively high gas permeability, sensitivity to oils, and poor odour resistance. It is less expensive than most films and is therefore widely used. High-density polythene (HDPE) is a stronger, thicker, less flexible and more brittle film, which has lower permeability to gases and moisture.

Traditional clay curd pots

Heat-sealing over a candle flame

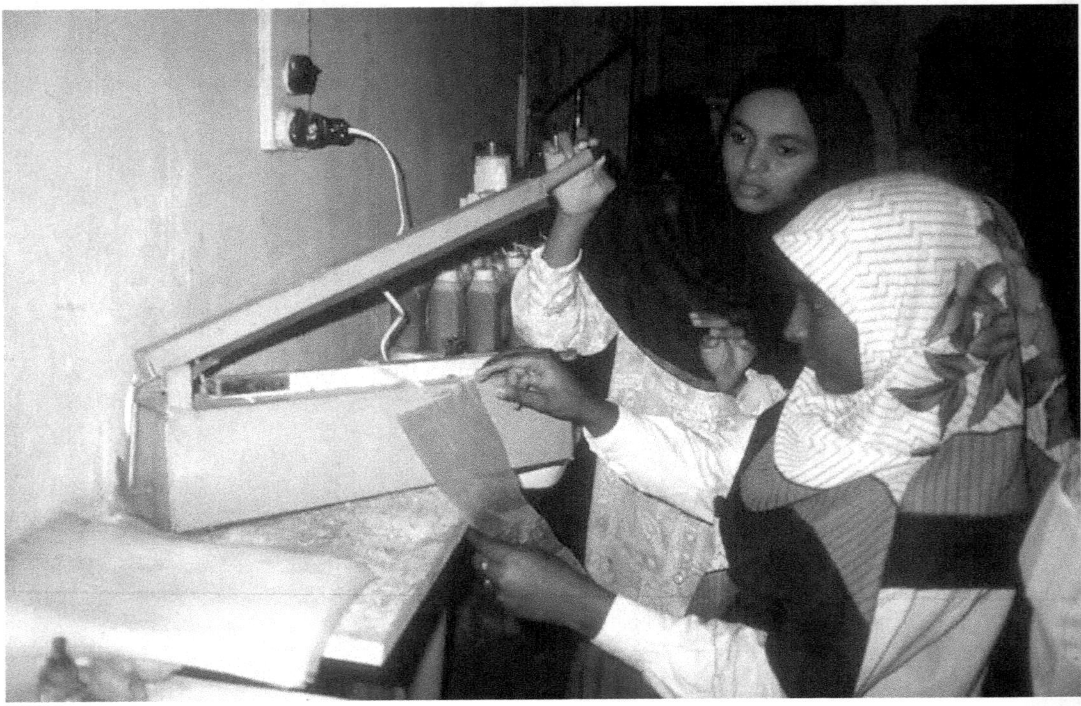

Heat-sealer machine

Polypropylene is a clear glossy film with a high strength and puncture resistance. It has low permeability to moisture, gases and odours, which is not affected by changes in humidity (as is the case with cellulose). It stretches, although less than polythene, and has good resistance to oil for packaging oily products.

Plain cellulose is a glossy transparent film which is odourless and tasteless. It is tough and resistant to puncturing, although it tears easily. However, it is not heat-sealable and the dimensions and the permeability of the film vary with changes in humidity. It is used for foods that do not require a complete moisture or gas barrier.

Coated films have a layer of another polymer or aluminium to improve their barrier properties or to enable them to be heat-sealed. Laminated films are two or more films glued together to improve the appearance, barrier properties or mechanical strength of a package. Aluminium foil is widely used in laminated films where low gas, water vapour, odour or light transmission is required.

Flexible films are usually made into bags and heat-sealed. Bags can either be supplied from a distributor or made on site using a form-fill-seal machine (Directory section 26.1). It is possible to seal bags by folding the edge of the film over the teeth of a used hacksaw blade and passing the folded edge through a flame. However, this is slow and the finished seal may not be reliable. Electric heat-sealers are a better option (Directory section 26.1).

Metal cans

Although aluminium and steel cans have a number of advantages over other types of container (e.g. they provide total protection of the contents and are tamperproof), they are not widely used in small-scale processing because of their relatively high cost and more complex and expensive sealing equipment (see Directory section 5.0).

PROCESSING OF FOOD GROUPS

Fruits and vegetables

Fruit and vegetables have similar compositions and share the same methods of cultivation, harvesting and storage. The main difference between them is their level of acidity. With a few exceptions, nearly all fruits are acidic and therefore called 'high-acid foods' (the acidity level of tropical fruits such as banana, melon and papaya is lower). In general, all vegetables are 'low-acid' foods that have a greater risk of transmitting food-poisoning bacteria (Part I) and this results in different processing methods and products.

Distinguishing between fruit and vegetables is difficult and often misleading. According to the botanical definition, a fruit is the portion of a plant that contains the seeds. Therefore, tomatoes, cucumbers, peppers and aubergines are strictly fruits but are usually referred to as vegetables. A distinction between fruits and vegetables can be made by their usage: plants that are eaten with a staple as a main course of a meal are considered to be vegetables, whereas those that are eaten as a dessert are fruits. This classification is used in this book.

Fruits

Fruits are commonly grouped according to their botanical structure, chemical composition and climatic requirements:

- berries are fruits which are generally small and fragile and grow in clusters
- drupes (stone fruits) contain single stones (including apricots, peaches, cherries and plums)
- pomes contain many seeds (including apples, quinces and pears)
- citrus fruits have a tough skin and have high levels of citric acid (oranges, lemons, limes and grapefruit)
- tropical fruits grow in warm climates (including bananas, dates, guavas, figs, pineapples, mangoes, jakfruit, breadfruit etc.).

Nutritional value

Fruits are largely composed of water (70–90 per cent of the edible portion), but they are also a valuable source of minerals, fibre and vitamins (especially vitamin C in citrus fruits and berries). Some vitamins are highly sensitive to light, temperature and air, and water-soluble vitamins are destroyed during processing, thus reducing the nutritional value of the fruit. The orange coloured fruits (mango, apricots, papaya and peaches) are all good sources of β-carotene, a precursor to vitamin A, which is needed to protect against infection and to prevent night blindness. Bananas are particularly high in potassium, a mineral that is essential for maintaining the correct water balance of the body. All fruits contain a high proportion of soluble fibre, which helps to maintain a healthy digestive system and may provide some protection against the development of bowel cancer.

Processing

The high acidity of fruits combined with the relative simplicity of processing makes them suitable for processing at the small scale. Traditionally, drying is the most appropriate form of processing fruits if the harvest season coincides with hot dry weather. Preservation with added sugar is another traditional preservation method, and fruit leathers, pastes, chutneys and jams are all popular products in some regions of the world. Practically any fruit can be processed, but some important factors which determine whether it is worthwhile are:

- the demand for a particular fruit in a processed form
- the quality of the raw material and whether it can withstand processing
- the regularity of supply of raw material.

Most fruits can be processed into higher value products

In many tropical countries, the succession of different fruits throughout the year means that fresh fruit is always available and there is no demand for processed products. In other situations the local varieties of fruit are not suitable for processing, because they lack flavour, are too fibrous or disintegrate during heat treatment or handling. Even if a fruit variety is suitable for processing, it may not be available in sufficient quantities to make processing an economical venture. Seasonal gluts may be processed for home use, but this cannot form the basis of a successful small business unless there is a regular supply or succession of different fruits for a large part of the year, or raw materials are part-processed and stored for future production.

A common problem for small producer groups is their lack of market research. Such enterprises are often supply-led and make products to use up a glut rather than to meet a defined market need for the processed product. This is not a sustainable business. Even if there is an identified market for processed fruits, there is usually competition from imported (often subsidized) products, and lack of packaging materials or professionally designed labels make products from small producers uncompetitive. Furthermore, although rural production reduces transport costs for raw materials, the markets may be a long way from the producers, which may cause difficulties in negotiations with retailers and high distribution costs, which further reduce competitiveness.

Another common mistake is to assume that poor quality fruits can be used up by processing. It is only possible to use substandard fruit if it has been rejected for cosmetic reasons (e.g. skin blemishes or the wrong size for the fresh market). Only high quality raw materials can produce high quality processed products.

These points are not meant to discourage anyone from starting such a venture, but the problems should not be underestimated. It is best to first seek advice from a qualified technical source. More detailed information on each of the above considerations is described by Fellows (1997a).

Fruit juices

Fruit juices are made from pure filtered fruit juice with nothing added. In some places sodium benzoate is added as a preservative to extend the shelf life, but this is not essential, and properly pasteurized juice has a shelf life of several months. Chemical preservatives cannot be used to cover up for poor hygiene or unsanitary conditions. Any fruit can be used to produce juice, but common ones include pineapple, orange, grapefruit, passion fruit and mango. Some juices, such as guava, are not filtered after pulping and are known as fruit nectars. Fruit juices rely on a combination of acidity, pasteurization and packaging in sealed containers for their preservation.

Process notes

Raw material preparation

Wash the incoming fruit with clean water. Chlorinate with bleach if necessary. Care is needed when handling pineapple as an enzyme in the juice damages skin after prolonged contact. Workers should therefore wear gloves to protect their hands. The juice must be heated to a higher temperature for a longer time to destroy the enzyme (e.g. boiling for 20 minutes).

Most fruits can be used to extract juice – grapefruit is a popular choice

Bottled fruit juices and squashes are popular products

Process flow sheet for fruit juice

Stage	*Equipment required*	*Directory section*
Select mature, undamaged fruits. Sort fruit, discard any mouldy or under-ripe fruits ↓		
Wash ↓		
Peel, de-stone and chop fruit or extract juice ↓	Peeler Juice extractor Pulper or blender Steamer Reamer (for citrus fruit)	29.0 31.0 31.0 2.1 31.0
Press ↓	Press	30.1
Filter juice to remove pieces of pulp ↓	Filter (bags) Sieve	18.0 18.2
Fill and seal ↓	Filler Sealing machine Capper Bottle washer	17.1 26.1 26.2 3.0
Heat-pasteurize in bottles ↓	Boiling water bath Double jacketed pan Plate heat exchanger Thermometer Clock	28.0 27.1 28.0 38.5
Cool rapidly to room temperature in cold water ↓		
Label ↓	Label gummer	26.5
Store away from sunlight in a cool place		

Pulping/juice extraction
Juice can be extracted in a number of ways:

- by steaming the fruit (especially for melon, papaya)
- by reaming (for citrus fruit)
- by pressing
- by pulping, using purpose-made pulpers or blenders (pineapple, mango etc.).

Fruit is crushed or pulped to extract the juice. For citrus fruits, a hand presser or a revolving citrus 'rose' can be used. Others require pulping using manual or powered pulpers/sieves, all of which force the fruit pulp through interchangeable metal strainers. Alternatively fruit can be pulped in a liquidizer, or soft fruits (papaya, melon etc.) can be steamed.

Pressing
A range of presses are available. Some larger machines are called filter presses, which carry out two processes – pressing followed by filtering to remove fruit particles.

Filtering
To produce clear juice, strain using a muslin cloth bag or stainless steel filter. Although juice is naturally cloudy, some consumers prefer a clear product. To achieve this it is necessary to remove the pectin haze by breaking down pectin using pectic enzymes. These can be obtained as commercially produced powders in some countries.

Filling
It is essential that the containers are thoroughly washed and sterilized. Small-scale production uses jugs and funnels, but for higher production rates a stainless steel bucket, drilled to accept an outlet tap, has proved to be a successful filler. Output can be doubled simply by fitting a second tap on the other side of the bucket. This system has been used to produce 500–600 bottles of juice per day in the West Indies. Small hand-operated or semi-automatic piston-fillers are also available. At a large scale, a different process is used: juice is pasteurized and then transferred into sterilized containers and then sealed. However, the higher costs and risk of contamination after pasteurization make this process less suitable for small-scale production.

Pasteurization
At a small scale, this may be carried out in a stainless steel, enamelled or aluminium pan over a gas flame, but this can result in localized overheating at the base of the pan, with consequent flavour changes. The next industrial jump in pasteurization is expensive and involves the purchase of a double-jacketed stainless-steel steam kettle and a small boiler. Larger-scale production uses a different technology (a plate heat exchanger), which is even more expensive.

Quality assurance

The main control points are as follows:

Fruit selection

All fruit should be ripe and free from bruising or insect damage and mould. It should be well washed and peeled and stoned before use. All rotten fruit should be removed, as this will spoil the final product. Fruit washing machines are available for large-scale production. Berries and soft fruits such as apricot and tomato are very fragile and should be handled carefully.

Filtering
To obtain clear bright juice it is necessary to filter the juice to remove fine suspended fruit particles.

Filling
Particular attention should be paid to the quality of reusable bottles, checking for cracks, chips etc. and washing thoroughly before using. Filtered juices are transferred into pre-sterilized bottles and sealed. Only new caps should be used and the correct fill-weight checked.

Pasteurization
The filled bottles are pasteurized by heating in boiling water for 5–10 minutes, depending

upon the size of the bottle. The temperature *and* time of heating are critical for achieving both the correct shelf life and retaining a good colour and flavour. A thermometer and clock are therefore needed.

Cooling

After heating, the bottles are cooled to room temperature by immersing in cold water, taking care not to break glass containers by cooling too rapidly.

Squashes and cordials

These consist of a 30 per cent mix of fruit pulp and sugar syrup, to give a final sugar level of about 12–14 per cent. All fruits contain sugar, usually around 8–10 per cent, and the addition of sugar to give the recommended levels must take into account the sugar already in the juice. However, in practice the amount of added sugar is usually decided by what the consumers actually want. In all cases, sugar syrups should be filtered through muslin cloth prior to mixing to remove particles of dirt which are always present.

The process is similar to the production of juice, except that sugar is added before filling containers. To avoid the use of large, expensive stainless steel pans, a large aluminium pan can be used to boil sugar syrup. A given amount of the syrup is then mixed with fruit juice in a small stainless steel pan, and this increases the temperature to 60–70 °C. The juice/syrup mixture is then quickly heated to pasteurizing temperature and hot-filled into bottles and sealed.

The Pearson square is a useful tool for calculating quantities of two solutions to mix to produce a final known concentration. See pp. 25–6 for examples of how to use the Pearson square.

Jams, jellies, marmalades and fruit cheeses

Collectively known as preserves, these products are increasing in importance in many countries, particularly in the more affluent urban areas. Fruits are most commonly used as the raw materials, but some vegetables such as pumpkin or jakfruit can also be used. The principles of preservation are heating to destroy enzymes and micro-organisms, combined with high acidity and sugar content to prevent re-contamination. The correct combination of acidity, sugar and the gelling compound 'pectin' (naturally present in most fruits or added in a commercially produced form) is needed to achieve the desired gel structure. The ingredients are then boiled together to evaporate water and achieve the correct solids content.

Jam is a solid gel made from the pulp of a single fruit or a combination of fruits. The fruit content should be at least 40 per cent, and in mixed fruit jams the first-named fruit should be at least 50 per cent of the total fruit added (based on European legislation). The total sugar content should not be less than 68 per cent to prevent mould growth after opening the jar, and in tropical climates 70 per cent is preferable. This can be measured using a refractometer (Directory section 38.4).

Jellies are crystal-clear jams, produced using filtered juice instead of fruit pulp.

Marmalades are mainly produced from clear citrus juices and have fine shreds of peel suspended in the gel. Commonly used fruits include limes, grapefruits, lemons and oranges. Ginger may be also used alone or in combination with the citrus fruit. The fruit content should not be less than 20 per cent citrus fruit, and the sugar content is similar to jam.

Fruit cheeses are highly boiled jam-like mixtures that have a final sugar content of 75–85 per cent, and thus set as a solid block. They can be cut into bars or cubes, or further processed as ingredients in confectionery or baked goods.

Raw materials and ingredients

For each of the above products, the fruit should be as fresh as possible and slightly under-ripe. In practice, a mixture of ripe and under-ripe fruit is satisfactory, but over-ripe fruit makes unsatisfactory jam as it is unable to set properly because of insufficient pectin

Jams, jellies and marmalades rely on their high sugar content for preservation

and/or acid. The richest sources of pectin are the peels of citrus fruits such as lime, lemon or orange and passion fruit or apple (known as 'pomace' after juice has been extracted). Fruits that are high in pectin may also be added to other fruits if there is insufficient natural pectin to form a gel. Commercially, pectin is available as either a light brown powder or as a dark liquid concentrate. It is stable if stored in a cool, dry place and it will lose only about 2 per cent of its gelling power per year. There are two main types of pectin:

- high methoxyl (HM) pectins that form gels in high solids jams (above 55 per cent solids) in a pH range of 2.0–3.5
- low methoxyl (LM) pectins, which do not need sugar or acid to form a gel, but instead use calcium salts.

LM pectins form a gel with a wide range of solids (10–80 per cent) within a broad pH range of 2.5–6.5. They are used mainly for spreads or for gelling agents in milk products. However, there are a large number of different types of pectin within each group, such as 'rapid set' and 'slow set' pectins, that are made for different applications, and it is necessary to specify carefully the type required when ordering from a supplier.

In addition to pectin, the fruit should have a high level of acidity to enable the gel to form. This is not a problem with most fruits, but melon, banana and papaya have relatively low levels of acid and it is necessary to add either citrus juice or commercially produced citric acid. Sugar has two main roles: it is needed to set the preserve and to prevent microbial spoilage. The final concentration must be high enough (>68 per cent) to prevent fermentation by yeasts or mould growth, but low enough (<72 per cent) to prevent crystallization.

Process flow sheet for jams and jellies

Stage	*Equipment required*	*Directory section*
Select mature, undamaged fruits. Prepare raw material ↓		
Peel, de-stone and chop fruit; remove all peel, stems, seeds and stones as necessary ↓	Peeler	29.0
	Juice extractor:	31.0
	steamer	2.1
	reamer (for citrus fruits)	31.0
	press	30.1
	pulper or blender	31.0
	Sieve/straining bag (for marmalade)	18.2/18.3
Weigh out and add sugar, pectin and citric acid. Check the pH (3.0–3.3) ↓	Scales	39.0
	pH meter (optional)	
Heat – slowly initially, then boil rapidly, avoiding localized burning until the final total sugar content (68%) is reached ↓	Boiling pans (stainless steel)	27.0
	Refractometer	38.4
	Thermometer	38.5
Fill (hot-fill at 82–85 °C into clean pre-sterilized jars and seal with a lid) ↓	Filler	17.1
	Sealing machine	26.1
Cool rapidly to room temperature in cold water ↓		
Label ↓	Label gummer	26.5
Store away from sunlight in a cool place		

Process notes

Raw material preparation

Wash the incoming fruit with clean water. Chlorinate with bleach if necessary.

Pulping/juice extraction
Juice can be extracted in a number of ways:

- by steaming the fruit (especially for melon, papaya)
- by reaming (for citrus fruit)
- by pressing
- by pulping, using purpose-made pulpers or blenders (pineapple, mango etc.).

Straining
To produce clear juice for the production of jellies and marmalade, strain using a muslin cloth bag. Additionally, sugar syrups should be strained to remove any unwanted material.

Calculation of total solids
The following recipes are only guidelines since they depend largely on the composition of fruit and the consumer tastes for sweetness, acidity and consistency. They assume that the fruits used are low in pectin – hence commercial pectin is added. The total yield is calculated according to the calculation at the foot of the page.

The yield for a product containing fruit:sugar in a 50:50 ratio, with desired solids in the finished product of 68 per cent, is 16.3546 kg, calculated as follows:

Component	*Soluble solids (kg)*
10 kg fruit at 10% TSS	1
10 kg sugar	10
60 g pectin (grade 200)	0.06
55 g citric acid	0.055
Total soluble solids (TSS)	11.115
Yield = 11.115 × 100/68	16.35

Weighing
Accurate scales are required for measurement of small amounts of ingredients. Pectin powder has to be mixed with sugar before it is added, to prevent lumps from forming.

Boiling
Use a stainless steel or enamelled metal pot; for larger production use a steam jacketed pan. There are two heating stages: initially, it is necessary to heat the fruit slowly to soften the flesh and extract the pectin. Once this is completed, it is vital to boil the mixture rapidly. This change in heat output is difficult to achieve without an easily controllable heat source, and gas and electricity are the preferred choices for most small commercial operations. During boiling avoid localized overheating which leads to burning and colour changes.

Boiling is carried out until the desired sugar content is reached. A refractometer is used to accurately assess sugar content. (The boiling temperature can also be used as a less accurate measure. The advantage is that a thermometer is cheaper than a refractometer.) Alternatively checks can be made by placing a drop of product in cold water to see if it sets. This is less accurate and requires experience and skill to work effectively.

Filling
It is essential that the containers are thoroughly washed and sterilized. The ideal temperature for pouring is 82–85 °C – hotter than this and condensation will form under the lid and drop down onto the surface of the preserve. This will dilute the sugar on the surface and allow mould growth. Colder than this the jam will begin to set and be difficult to pour, and a partial vacuum will not form in the jar. Jars should be filled to about 9/10 of their volume. Small-scale production uses jugs and simple funnels, but for higher production rates small hand-operated or

$$\frac{\text{Total soluble solids (TSS) content of raw ingredients} \times 100}{\text{Percentage of total soluble solids in final product}}$$

semi-automatic piston-fillers are available. Jars are held upright while the gel forms during cooling.

Packaging

Ideally, glass jars should be used, with new metal lids. It is possible to use paper, polythene or cloth tied with an elastic band or cotton to cover the jars, but the appearance is less professional, and there is a risk of contamination by insects. The use of paper is not recommended unless metal lids are impossible to obtain. Increasingly, preserves are being packaged in plastic pots with aluminium foil lids as they are cheaper and more convenient than glass. Alternatives include plastic pouches or sachets (see Packaging). Technical advice should be sought if these packs are being considered.

Quality assurance

The main points are as follows:

Fruit selection

All fruit should be ripe and free from bruising or insect damage and mould. It should be well washed and peeled and stoned before use.

Ingredient mixing

Accurate scales are required to ensure that the correct weights are used each time. Two sets of weighing scale will be required: one with a large capacity for the fruit and sugar, and a smaller set for pectin.

Acidity

Acid (usually citric acid, but also tartaric acid or malic acid) is added if the natural fruit acids are not sufficient to ensure a pH of 3.0–3.3.

Sugar

Refined granular sugar should be used to produce a high quality preserve. Often it contains impurities and it is advisable to dissolve the sugar in water to produce a strong syrup that is filtered through a fine mesh prior to use.

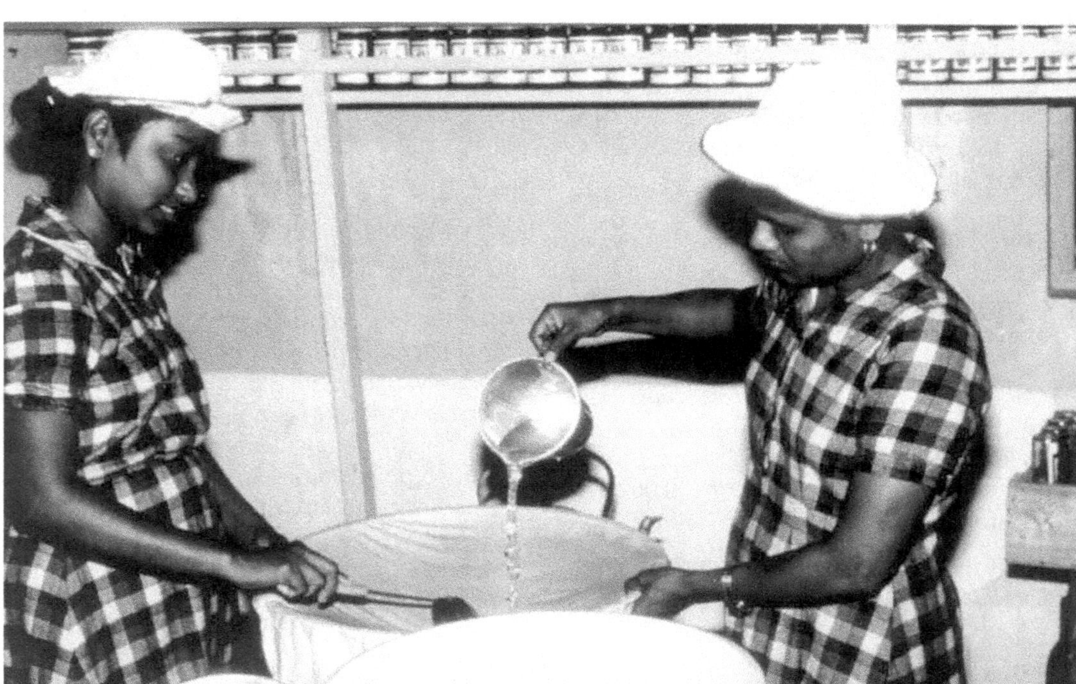

Filtering sugar for jam making results in a clearer jam or jelly

Pectin
Slow setting pectin is for preserves that are left to cool and set in the jars. Fast setting is needed for larger containers or for preserves that contain suspended pieces of fruit or fruit peel. The concentration varies from 0.2–0.7 per cent depending upon the type of fruit being used. Pectin is usually supplied as 150 grade (or 150 SAG) which indicates the ratio of the weight of sugar to pectin needed to obtain a standard gel at 65 per cent soluble solids: 5 SAG is normally enough to produce a good gel, thus the 150 SAG pectin needs to be diluted 30 times to produce a 5 SAG strength. Therefore, 3.3 g 150 SAG pectin would be used for every 100 g of sugar. If commercial pectin is not available, it is possible to produce a pectin solution by boiling sliced skins of passion fruit or citrus fruit in water for 20–30 minutes (see page 52). The solution should be filtered before it is added to the fruit pulp. The amount needed depends on the type of fruit and the concentration of the pectin solution, and can only be found by trial and error.

Dried fruits

The principles of drying are covered in more detail in Part I. Drying fruits in the sun is simple and has the advantage of low or no fuel and equipment costs. It is the method used for raisin and sultana production, the largest volume dried fruits, and also for dates. However, for other types of fruit the change in colour is unacceptable and a different method is required. Solar drying and the use of an artificial dryer are both possible options, and the use of sulphur dioxide is another (see Part I). Sulphur dioxide improves the colour and increases the shelf life of some dried fruits. There are two methods of applying sulphur dioxide: sulphuring and sulphiting. Sulphiting involves soaking the fruit in a solution of sodium sulphite or sodium metabisulphite, and sulphuring is achieved by burning sulphur in a sulphur cabinet. The strength of the sulphite solution or the amount of sulphur used and the time of exposure depend on the commodity, its moisture content and the levels permitted in the final product which are set by legal standards in each country. Typically a 3 g/l sulphite solution or 2 g of sulphur for each kilogram of prepared fruits is used with a sulphuring time of 60–90 minutes. (NB: importers in the European Union and United States may specify that sulphur dioxide is not to be used in products.) An example of the production of a dried fruit follows.

Dried mango slices

Mangoes are washed, peeled and cut into 6–8 mm thick slices with a stainless steel knife. The slices are soaked for 18 hours in a solution containing:

boiling water	1 litre
sugar	700–800 g
potassium metabisulphite	3 g
lemon juice	20 ml.

The slices are drained and placed on aluminium trays coated in cooking oil or glycerine, which are placed in the sun or in a solar/artificial dryer. Drying is completed when the product has a moisture content of 15 per cent (see Part I for methods of assessing moisture content). The dried slices are packed in cellophane bags, labelled and stored in a dry place. If correctly packaged and stored, they have a shelf life of up to nine months.

Fruit leathers

Fruit leathers are dried sheets of fruit pulp which have a soft, rubbery texture and a sweet taste, with the distinctive flavour and colour of the raw material. Most fruits can be used to make leathers, including banana, mango, apricot, guava, pineapple and papaya. The leathers can be rolled in plastic sheets and stored for a few weeks, then cut into small pieces for use in confectionery and baked goods. Layers of fruit leather from different raw materials can be sandwiched together to form a multi-layered confectionery product. The process involves the production of a fruit pulp which is then dried. It is important to select ripe fruits, but

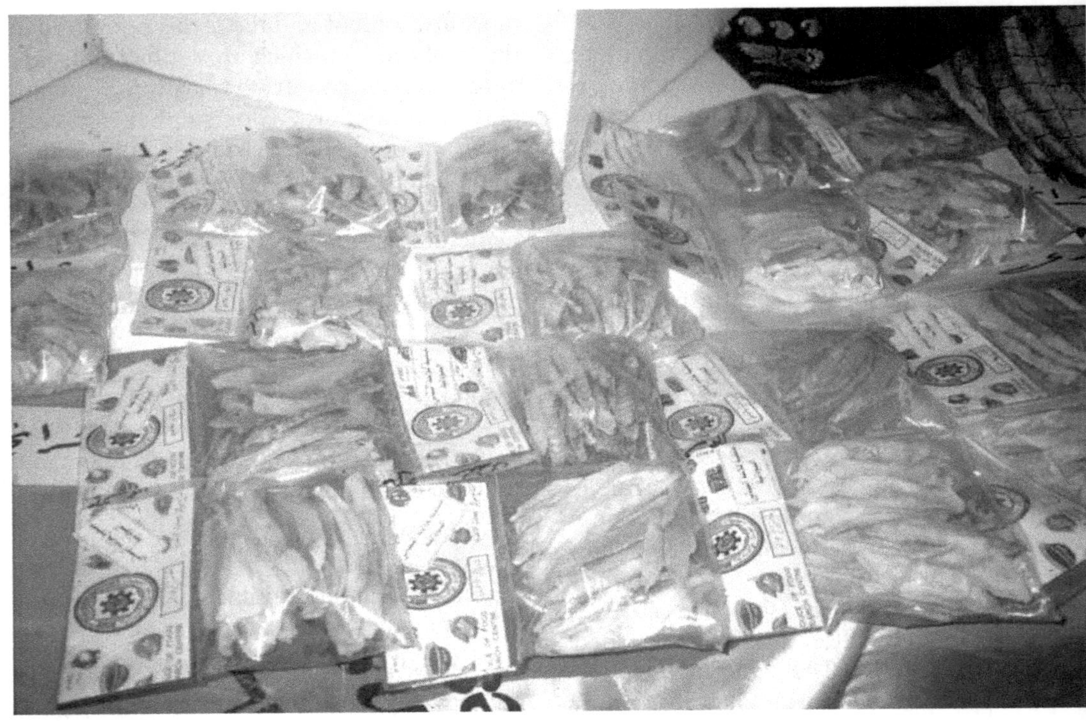

Dried mangoes ready for sale

over-ripe fruit can easily become damaged and bruised. Under-ripe fruits do not have a full flavour and colour. If bananas are used, they should be harvested before they are fully ripe, with the skin a little green. Fruits are pulped and then spread onto polythene sheets for drying at 38–60 °C. The dried product is packaged in moisture-proof plastic and stored in a cool, dry place, away from sunlight.

Osmotically dried fruit

This product is similar to dried fruit, but with a sweeter taste and a softer texture. The colour is often better retained than with simple drying. Fruits are peeled and cut into small pieces. Dipping in citric acid for 5–10 minutes or sulphuring can be used to retain a brighter colour, but sulphuring should not be used for red fruits as it bleaches the colour. Sulphuring can either be done by burning sulphur – 350–400 g/kg fruit (2–3 tablespoons per kg fruit for 20–40 minutes) – in a sulphuring tent or cabinet or dipping in sulphite solution (0.3–0.45 g sodium or potassium metabisulphite in 1 litre of water).

After sulphuring, the fruit pieces are boiled in a strong sugar solution (60–70 per cent sugar) for 10–15 minutes, then soaked in the syrup for up to 18 hours. After soaking, the fruit is dried in the normal way.

Fermented fruit products

Fruits are used in the production of a range of fermented products, including pickles, vinegar, wine and banana beer.

Wines and beers

Grape wine is the most economically important fruit juice alcohol, but wines are made commercially from a wide range of fruits including pineapple, mango, passion fruit and jakfruit. Banana beer is a traditional product that is widely produced in Eastern and Southern Africa.

RED GRAPE WINE

This is an alcoholic drink containing 10–14 per cent alcohol made from the fruit of the grape plant (*Vitis vinifera*). The colour ranges from light to deep red. Many varieties of grape are used, including Cabernet Sauvignon, Grenache, Nebbiolo, Pinot Noir and Torrontes (Ranken et al., 1997). The grape skins contain tannins, which contribute to the flavour, and alcohol produced during the fermentation assists with extraction of pigments from the skins to give the final colour to the wine.

Processing

Ripe and undamaged grapes are crushed to yield the juice plus skins (known as 'must'), which is transferred to fermentation vessels. The initial fermentation takes between one day and three weeks depending on the required colour of the final product. The skins are then removed and the partially fermented wine is transferred to a separate tank to complete the fermentation over several weeks. The fermentation can be from naturally occurring yeasts on the skin of the grape, or by using a starter culture of *Saccharomyces cerevisiae*. If the fermentation is allowed to proceed naturally, the end result is less controllable, but it produces wines with a different range of flavour characteristics (Rhodes and Fletcher, 1966; Colquichagua, 1994; Fleet, 1998).

Traditionally, fermentation was carried out in wooden barrels or concrete tanks, but modern wineries now use stainless steel tanks as they are more hygienic and enable better temperature control. Fermentation stops naturally when all fermentable sugars have been converted to alcohol, or when the alcohol strength reaches the limit of tolerance of the strain of yeast involved. The product is packaged in coloured glass bottles to reduce light transmission to the wine, and sealed with corks made from the bark of the cork oak (*Quercus suber*), or a synthetic plastic 'cork', to exclude air.

Fruits form the ideal substrate for fermentation

BANANA BEER

Banana beer is an alcoholic beverage produced from specifically grown 'beer bananas' (*Musa* species) mixed with a cereal flour (often sorghum or millet flour) and water and then fermented. It is sweet and slightly hazy, with a shelf life of several days under correct storage conditions.

Processing

The extraction of a high yield of banana juice, without excessive browning or contamination by spoilage micro-organisms, and filtration to produce a clear product are of great importance. Ripe bananas are peeled and squeezed with grass to cut through the cells of the banana so that only a clear juice is obtained. The residual pulp remains in the grass. One part water is added to every three parts banana juice to reduce the sugar levels and allow the yeast to act. Cereals are ground and roasted to improve the colour and flavour of the final product. The mixture is placed in a covered container and fermented for 18–24 hours (Fellows, 1997b). It is essential that proper hygiene procedures are followed to prevent contaminating bacteria from competing with the yeast and producing acid instead of alcohol. After fermentation the product is filtered through cotton cloth and bottled for sale. Packaging is required to keep the product for its relatively short shelf-life, and clean glass or plastic bottles are used. The product should be kept in a cool place away from direct sunlight.

Fruit vinegar

Vinegar can be made from almost any fermentable material (e.g. fruits, waste fruit skins, honey, syrups, cereals, beer or wine). Whatever the raw material used, the fermentation process is a sequence of yeast followed by acetic acid bacteria (*Acetobacter pasteurianus*). Yeasts ferment sugars to alcohol and acetic acid bacteria oxidize the alcohol to acetic acid. For a good fermentation, an alcohol concentration of 10–13 per cent is optimum. If it is much higher the alcohol is not completely oxidized to acetic acid, and if it is lower than 13 per cent there is a loss of vinegar yield because the acetic acid itself is oxidized.

Vinegar can be made quite simply by the spontaneous fermentation of wine. However, this process is very slow and the vinegar produced by this method tends to be of inferior quality. Controlled fermentation conditions produce a more acceptable product.

PINEAPPLE PEEL VINEGAR

Vinegar from waste pineapple peels has a distinct, light colour and pineapple flavour. The peels (not the leaves or stems) should be from washed ripe pineapples. Damaged, rotten or infected fruits should not be used. The peels are liquidized and diluted with water (water:pulp ratio of 4:1). The pH is adjusted to 4.0 using sodium bicarbonate, and yeast nutrient (ammonium phosphate) is added at 0.14 g per litre. Sugar and clean water may also be added, and the liquor is put into plastic or clay pots (aluminium or iron pots should not be used since the acid in the pineapple reacts with the metal). Each pot is then inoculated with a yeast starter culture at 2.7 g per litre and covered with a clean cotton cloth to prevent contamination by insects or dust. The fermentation is allowed to take place at 25 °C for two days. The 'must' is then filtered and inoculated with acetic acid bacteria and allowed to ferment for eleven days with aeration of the must in a special aerated fermenter (Directory section 43.0). The product should become increasingly acidic, and by the eighth day it should have the required concentration of 4–5 per cent acetic acid. If higher acidity is desired the product is left to ferment for another one or two days. The development of acidity should be checked by tasting the product during fermentation. The fermented vinegar is strained through a cheese-cloth. The residual bacteria, which are retained in the sediment in the cheese-cloth, can be saved and reused as a starter culture two or three more times. The filtered vinegar is bottled in clean sterilized glass bottles with lids that are loosely closed. The filled bottles can be immersed in a water bath and heated to boiling (100°C) for about 10 minutes to pasteurize the product and improve its stability. This step is not essential, but does improve the shelf life of the product. The pasteurized bottles are allowed to cool and the bottle caps are tightened during cooling. The bottles should be stored in a cool, dark place.

Utilizing fruit waste

Processing of fruits produces two types of waste: a solid waste of peel/skin, seeds, stones etc., and a liquid waste of juice and washwaters. In some fruits the discarded portion can be very high (e.g. mango 30–50

Production of vinegar is one way of utilizing waste fruit and peels

per cent, banana 20 per cent, pineapple 40–50 per cent and orange 30–50 per cent) and can produce a serious waste disposal problem. This can lead to problems with flies and rats around the processing room if not correctly dealt with. If there are no plans to use the waste it should be buried or fed to animals well away from the processing site.

There are a number of possibilities for using solid fruit wastes but as yet there is no evidence that any of these are economically viable. It is stressed that a full financial evaluation should be done before introducing any of the ideas below.

One of the main problems in using fruit waste is to ensure that it has a reasonable microbiological quality, and processors should only use wastes on the day that they are produced. Even with this precaution, waste is still likely to contain mouldy fruit discarded during processing together with insects, leaves, stems and soils. These will contaminate any products made from it unless some preliminary separation takes place during processing (e.g. peel and waste pulp into one bin, mouldy parts, leaves, stones, seeds, soil etc. in another, which is discarded).

Candied peel

Peel from citrus fruits (orange, lemon, lime, grapefruit) can be candied for use in baked goods or as snack foods, and shreds of peel are used in marmalades. The process to make these involves cutting the peels to fine shreds and boiling them in a 20 per cent sugar syrup for 15–20 minutes. The sugar concentration in the syrup is then progressively increased to 65–70° Brix (percentage sugar measured by a refractometer – Directory section 38.4) as the shreds are soaked for 4–5 days. They are then removed, rinsed and dried in the sun or in a dryer (Directory section 14.0). Candied peel can form a second product for a fruit juice or jam processor, especially if larger food companies are willing to buy the peel as an ingredient for their foods.

Oils

The stones of some fruits (e.g. mango, apricot, peach) contain appreciable quantities of oil or fat, some of which have markets for culinary or perfumery/toiletry applications. In addition some seeds (e.g. grape, papaya and passion fruit) contain oils which have very specialized and valuable markets. The main problems are to identify the import/export agents who would buy such products, producing the oil in sufficient quantities to meet their orders, meeting their very stringent quality standards and finally obtaining the equipment needed to produce the oils at low cost.

The process involves grinding the seeds/nuts in a powered hammer mill (Directory section 23.3) to release the oil without a significant rise in temperature, which would spoil their delicate flavours. A press is needed to extract the oil, but to our knowledge the existing manual presses have not been tried in this application, and experimentation is needed to establish oil yields and suitability of the equipment. Alternatively steam distillation (Directory section 43.0) is well established for citrus peel oils at a small scale and could be adapted to other oils (the large-scale process of solvent extraction is not suitable for small-scale applications). The crude oil could be sold for refining elsewhere, but it is likely that at least preliminary refining would need to be carried out by the producer. It is also possible that the sale of seeds or stones to larger oil processors could generate additional income for small-scale fruit processors.

Pectin

Pectin is extracted commercially from citrus peel and apple pomace. Passion fruit is also a good source of pectin. In most developing countries pectin is imported and, superficially at least, there would seem to be a good market for supplying local fruit processors with pectin to substitute for imports. However, there are major problems:

- In countries where this has been tried, it has not been possible to produce pectin at a cost which is lower than the imported products.
- It is difficult to produce pectin powder on a small scale, though liquid pectin concentrate is possible.
- There are many types of pectin, each of which has specific properties that make it suitable for its intended application. The process to standardize the different types may be too difficult at a small scale, and a detailed knowledge of pectin and its properties is needed to ensure that a producer is supplying the right product.

However, the process of pectin extraction is not complex: shredded fruit peel or pomace is soaked in hot water (60–70 °C), or the hot water is recirculated through the material, and the pectin is extracted into the water along with sugars and other fruit components. This is continued until the pectin concentration increases to around 5 per cent. The pectin is then precipitated as a gel from the solution by adding one of a number of chemicals – the most common being hexane or spirit alcohol. After precipitation the solvent is recovered by distillation and reused (the percentage recovery and cost of this step are the most critical in determining profitability). It is also possible to use ammonium sulphate (a component of fertilizer) but this cannot be recovered, and the higher cost therefore prevents its use commercially in large-scale operations.

The pectin gel is then washed and redissolved in water to produce a concentrated pectin solution. It is at this stage that it is standardized or modified to give the specific properties required. On a large scale it is usually dried to a powder, but on a small scale it is possible to add sodium benzoate preservative and sell the concentrated liquid in bottles.

Reformed fruit pieces

Fruit pulp can be recovered and formed into synthetic fruit pieces. It is a relatively simple process but the demand for this product is not likely to be high, and a thorough evaluation of the potential market is strongly recommended before any work is undertaken. The process involves boiling fruit pulp to concentrate and sterilize it. Sugar may also be added, and a gelling agent (sodium alginate) is mixed with the cooled pulp. The mixture is fed as droplets into a bath containing a strong solution of calcium chloride where the calcium and the alginate combine to form a solid gel structure. This forms small grains of reformed fruit which can be used in baked goods. Alternatively the mixture can be poured into fruit-shaped moulds and allowed to set. The most common product of this type is glacé cherries.

Enzymes

Commercially, the three most important enzymes from fruit are papain (from papaya), bromelin (from pineapple) and ficin (from figs). Each is a protein-degrading enzyme, which is used in applications such as meat tenderizers, washing powders, leather tanning and beer brewing. However, it is unlikely to be economical to obtain these from waste fruit. Even the more efficient collection from fresh whole fruit is no longer economical, and changes in large-scale production (higher quality standards and use of biotechnology to produce 'synthetic' enzymes) mean that small-scale producers will be unlikely to compete effectively. However, recent moves away from synthetic ingredients and increased demand for organic ingredients have stimulated an increase in natural papain production, and further studies of this market may be worthwhile.

In summary, each of the above uses for fruit waste requires:

- good knowledge of the potential market for products and the quality standards required
- careful assessment of the economics of production
- a certain amount of additional knowledge
- additional capital investment in equipment
- large amounts of waste to make utilization worthwhile.

At small scales of operation, where localized pollution is more important than process economics, the most likely solution is to use fruit waste as animal feeds.

Vegetables

Vegetables are considered to be plants that are eaten with the main course of the meal. Vegetables can be grouped together according to the part of the plant from which they are taken (Table 6). Root vegetables (cassava, potato, yams etc.) are described in the next chapter.

Table 6. Vegetable groups (from Dauthy, 1995)

Category	*Examples*
Modified buds – bulbs	Onion, garlic
Leaves	Cabbage, spinach
Petioles (stalks)	Celery
Flower buds	Cauliflower
Sprouts, shoots	Asparagus, bamboo shoots
Legumes	Peas, green beans
Cereals	Sweet corn
Vine fruits	Squash, cucumber, pumpkin
Berry fruits	Tomato, aubergine (eggplant)
Tree fruits	Avocado, breadfruit, jakfruit

Vegetables are 'low acid foods' which provide an ideal substrate for food-poisoning micro-organisms (see Part I). They can be safely preserved by making them more acidic by pickling, or by salting, drying or pasteurization.

Nutritional value

Vegetables contain a high proportion of water (70–90 per cent), are low in fat and protein, have varying levels of carbohydrate and fibre, and have useful amounts of minerals and vitamins. Legume vegetables such as peas and beans contain higher levels of protein and are a good source of B vitamins. Green leafy vegetables contain high levels of vitamin C, β-carotene, folic acid, iron and other minerals. They are especially valuable when consumed raw as processing destroys much of the vitamin C. The orange/yellow vegetables (carrot, pumpkin, sweet potato and tomato) are all good sources of β-carotene which is converted by the body into vitamin A. All vegetables contain a high proportion of soluble and insoluble fibre, which helps to maintain a healthy digestive system and may provide some protection against the development of bowel cancer.

Processing

The similarities in composition between fruit and vegetables mean that traditionally they have been processed using similar methods and technologies. Sun-drying, addition of salt (instead of sugar in fruit processing) and fermentation are the most common methods used to preserve vegetables.

Salting and lactic acid fermentations

There are three basic types of process: dry salting, brining and non-salted fermentation. Salting provides a suitable environment for lactic acid bacteria to grow, producing the acid to preserve the vegetable and impart characteristic flavours and textures to vegetable products.

There are two methods of pickle production:

- lactic acid fermentation, either with or without the addition of salt
- preservation of vegetables in acetic acid (vinegar).

Salt for pickling

For pickling, any type of common salt is suitable as long as it is pure. However, salt that contains chemicals to reduce caking should not be used as the chemicals make the brine cloudy; salt with lime impurities can reduce the acidity and the shelf life of the product; salt with iron impurities can result in the blackening of the vegetables; salt having magnesium impurities imparts a bitter taste and salt containing carbonates can result in pickles with a soft texture (Lal et al., 1986).

Fermented pickles can be made by dry salting, by placing vegetables in a weak brine solution or by allowing the vegetables to ferment without salt.

DRY SALTED PICKLES

In dry salting, the salt extracts juice from the vegetables and creates a brine. The vegetables are placed in a layer about 2.5 cm deep in a barrel and a layer of salt is sprinkled over them. Alternate layers of vegetables and salt are then built up until the container is three quarters full. A cloth is then placed on the vegetables and weighted down for 24 hours to compress the vegetables and assist in the formation of a brine. As soon as the brine is formed, fermentation starts and bubbles of carbon dioxide appear. Fermentation takes between one and four weeks depending on the ambient temperature, and is completed when no more bubbles appear. The pickle can then be packaged with either vinegar and spices, or oil and spices (Lal et al., 1986). In another method 50 per cent salt and 50 per cent vegetables are used to create a saturated brine that inhibits fermentation but is high enough to preserve the vegetables.

Lime pickle

Dry salted lime pickle is particularly popular

in India, Pakistan and North Africa. It is brownish red and the lime peels are yellow or pale green with a sour and salty taste. It is eaten as a condiment with curries or other main meals. Fully ripe limes are dipped in hot water (60–65 °C) for about five minutes and cut into quarters, or four slits are made on the skin. They are placed in a layer of about 2.5 cm depth in a fermenting barrel, and dry salt is added in alternate layers (4 kg limes:1 kg salt) which extracts the juice and creates the brine. Spices, especially chillies, are added depending on local preferences. The container is then placed in the sun for a week, and fermentation starts as soon as the brine is formed. It is then moved to the shade and fermentation continues for a further one to four weeks depending on the ambient temperature.

Sauerkraut

Shredded cabbage is placed in a jar and salt is added. The function of salt is to withdraw juice from the cabbage so making a more favourable environment for the desired bacteria to develop. The amount of salt is important in the process and affects the quality of the final product. Too much salt may inhibit the desirable bacteria, although it may also contribute to the firmness of the sauerkraut. Dry salt is added at the rate of 1–1.5 kg per 50 kg cabbage (2–3 per cent).

The first sequence of lactic micro-organisms produce lactic acid, and when the acidity reaches 0.25–0.3 per cent these bacteria slow down and begin to die off. Then other species of lactobacilli take over and further raise the acidity to 2–2.5 per cent, which completes the fermentation. The acidity helps to control the growth of spoilage micro-organisms and contributes to the extended shelf life of the product. Changes in the sequence of desirable bacteria, or the presence of non-desirable bacteria, result in alterations to the taste and quality of the product. The optimum temperature for sauerkraut fermentation is around 21 °C and a variation of just a few degrees alters the microbial activity and affects the quality of the final product. Temperature control is therefore one of the most important factors in the sauerkraut process. Starter cultures ensure consistency between batches, inhibit undesirable micro-organisms and speed up the fermentation process because there is no time lag while the cells multiply. It is also possible to use the juice from a previous fermentation as a starter culture for subsequent fermentations.

BRINE FERMENTED VEGETABLES

A 10–15 per cent salt solution is prepared (as a guide, a fresh egg floats in a 10 per cent brine solution – Kordylas, 1990). Vegetables are submerged in the brine using weights to hold them under the solution, and the container is sealed. The strong brine draws sugar and water out of the vegetables, which decreases the salt concentration and permits rapid development of micro-organisms in the brine. Extra salt is added periodically to the brine to maintain the level. The rate of fermentation can be controlled by adjusting the concentration of salt and the temperature of the brine. The aim is to enable selection of salt-tolerant lactic acid bacteria by preventing yeasts, moulds and other spoilage micro-organisms until the lactic acid bacteria have grown to sufficient numbers. Then the production of lactic acid inhibits further spoilage.

Pickled cucumbers

Washed cucumbers are placed in large tanks and a solution of 6–15 per cent brine is added. To prevent spoilage, the cucumbers are submerged with weights, ensuring that none float on the surface. During fermentation, visible changes take place which are important in judging the progress of the process: the colour of the cucumber surface changes from bright green to a dark olive green as acids interact with the chlorophyll; the interior of the cucumber changes from white to a waxy translucent shade as air is forced out of the cells; and the density of the cucumbers increases as a result of the absorption of salt and removal of air – and

Pickle is a popular condiment, made from fruit and vegetables

they sink in the brine rather than floating on the surface.

Green mango pickle

The best pickles are obtained from fruit at early maturity when it has reached almost maximum size. Riper fruit results in pickles with a fruity odour and lacking the characteristic and predominant mango flavour. After washing, the fruit is cut with a single stroke to ensure minimum damage and avoiding mushiness in the final product. The sliced mangoes are soaked in brine (saturated salt solution). Sodium metabisulphite (1000 ppm) can be added as a preservative and 1 per cent calcium chloride can be added to maintain the firmness of the fruit. After fermentation, the brine is drained off and spices are mixed with the mango slices. The mixture is packed and oil added on the surface to exclude air.

NON-SALTED, LACTIC-ACID FERMENTED VEGETABLES

Some vegetables are fermented by lactic acid bacteria without the addition of salt. Examples include *gundruk*, *sinki* and other wilted fermented leaves. The fermentation relies on rapid colonization of the food by lactic acid bacteria, which lower the pH and prevent the growth of spoilage organisms. Shredded leaves are wilted and tightly packed in an earthenware pot, and warm water (at about 30 °C) is added to cover the leaves. The pot is then kept in a warm place (about 18 °C) for five to seven days, until a mild acidic taste indicates the end of fermentation. The *gundruk* is then removed and sun-dried.

NON-FERMENTED PICKLES

Atchar is a hot and spicy vegetable pickle commonly consumed in Asia. Different vegetables can be used according to preference and local availability, but a typical recipe is as follows.

Ingredients	
carrots	750 g
cabbage	600 g
green pepper (capsicum)	100 g
onions	450 g

cayenne pepper	120 g
ginger powder	15 g
salt	40 g
curry powder	30 g
sunflower oil	750 ml
vinegar	300 ml

Method

The vegetables are washed, peeled and chopped or grated to a uniform size, lightly cooked for about 5 minutes and mixed together with soft-fried onions, a blend of fried spices, salt and vinegar. It is heated for a further 5 minutes with stirring. The salt and vinegar act as preservatives, and it is essential that the ratio of these two ingredients is controlled to prevent spoilage through fermentation or mould growth. The final level of acetic acid should be 6–10 per cent.

Dried vegetables

Most vegetables can be dried, but the conditions for drying vary for different vegetable types. Most vegetables are blanched to stop enzyme activity, and speed up drying by softening the plant cells, which allows water to escape more easily. Without blanching, enzyme action continues during drying, there is a loss of colour and flavour, and a reduction in quality. However, blanching also destroys some water-soluble vitamins and causes minerals to be leached out. Tomatoes, okra, peppers, onions and mushrooms do not require blanching prior to drying.

Blanching can be carried out by placing cut vegetables in boiling water or by steaming for a few minutes. The time depends on the size of the vegetable pieces and the type of vegetable (see Table 7). Although steam blanching takes longer, it is the preferred method since it minimizes losses of vitamins and minerals. Cut vegetables are placed in a wire container or colander that is suspended over a pan of boiling water, and the pan is tightly covered.

Gundruk is made from fermented and dried leaves

Most vegetables – such as these tomatoes – can be dried

Table 7. Blanching times of selected vegetables (from Kordylas, 1990)

Vegetable	*Time for steam blanching (minutes)*
Aubergine (eggplant)	3.5
Leafy green vegetables	2–2.5
Green beans	2–2.5
Corn	2–2.5
Carrots	3–3.5
Cabbage	2.5–3
Cauliflower	4–5

Drying is described in Part I.

Chutneys and sauces

The basic principle of preservation for these products is the use of acetic acid to preserve the product by inhibiting the growth of spoilage and food-poisoning micro-organisms. Other ingredients such as salt and sugar add to the preservative effect.

CHUTNEYS

These are jam-like mixtures which have added vinegar and spices. The high sugar content exerts a preservative effect, and a high level of vinegar addition is not always needed. They are produced in a similar process to jams.

SAUCES

These are thick viscous liquids made from pulped fruit and/or vegetables. They are pasteurized to remove spoilage micro-organisms, and the addition of salt, sugar and vinegar preserves the product after opening. The flow sheet below outlines typical processing stages for a representative product, tomato sauce. The following ingredients are used:

tomatoes	20 kg
sugar	1.5 kg
onions	450 g, finely chopped
salt	330 g
vinegar	800 g
spices:	

mace	3.5 g
cinnamon	9 g
cardamom	11.25 g
cumin	11.25 g
ground black pepper	11.25 g
ground white pepper	5 g
ground ginger	5 g

A tomato chilli sauce, which is popular in many countries, can be prepared by adding 2.5 g chilli powder to 10 kg tomato pulp before processing.

Process notes

Preparation of raw materials

Select good quality, fully ripe red fruits, preferably 'plum' varieties without infection, mould or rot. Blanch in hot water for 3–5 minutes until the skin is loosened and remove skin. Chop or pulp either using a hand grinder or a pulper, depending upon the size of operation.

Mixing ingredients

Tie spices in a small muslin bag, add to the tomatoes with 500 g sugar and chopped onions.

It is necessary to calculate a value known as the 'preservation index'. This is used to assess whether the product is safe from food spoilage and poisoning micro-organisms. The value can be calculated as follows:

$$\frac{\text{total acidity} \times 100}{(100\text{-total solids})} = \text{not less than 3.6 per cent}$$

Heating

Heat to below boiling point in a heavy pan with continuous stirring until the mixture has reduced to half the original volume, remove the spice bag and add sugar, salt and vinegar. Continue heating for 5–10 minutes. Check the final total soluble solids (10–12 per cent solids) with a refractometer.

Filling

Hot-fill pre-sterilized bottles or jars at not

Preparation of tomato pulp for sauce and puree

Process flow sheet for tomato sauce

Stage	*Equipment required*	*Directory section*
Sort tomatoes ↓		
Wash ↓	Fruit cleaner	7.1
Heat ↓		
Pulp ↓	Fruit/vegetable chopper Cutting, slicing and dicing equipment Pulper	10.1 10.1 31.0
Mix ingredients (spices, onions, sugar) ↓	Weighing and measuring equipment	39.0
Heat ↓	Tilting boiling pan (stainless steel) Heat source Thermometer Refractometer	27.1 25.0 38.5 38.4
Separate spice bag ↓		
Mix ingredients ↓		
Heat ↓		
Fill/seal ↓	Filling machine Capping machine Sealing machine Bottle sterilizer	17.1 26.2 26.1 3.0
Cool ↓	Water bath	
Label ↓	Labelling machine	26.5
Store		

less than 80 °C, close the lids tightly and cool to room temperature. In this way the hot mixture will form a partial vacuum in the jar and help prevent re-contamination.

Label
Record the product name, date of manufacture, contents, brand name and name of manufacturer.

Store
If adequately packaged and stored in a cool place, the sauce can be stored for up to a year without any loss of flavour or taste. However, it should be stored out of direct sunlight to avoid any discoloration.

Quality assurance

The main points are as follows:

Raw material

Tomatoes should be fully ripe, sound and of the correct processing varieties with high solids content. Spices must be clean, mature and not infected with mould.

Batch formulation
The ratio of ingredients must be carefully monitored and controlled according to the formulation for each batch to produce the desired taste, consistency and flavour each time.

Heating
The product must be stirred continuously during heating to prevent burning. Ideally, a double pan heater should be used as this prevents localized heating and burning.

End-point
The end of boiling should be determined by checking the soluble solids content with a refractometer (10–12 per cent solids).

Root crops

Starchy root crops and tubers, including cassava, yam, sweet potato and Irish potato, provide cheap and readily available sources of calories for much of the world's population. Cassava and sweet potato are the most important root crops in the developing world, with a combined total annual production of around 300 million tonnes. They are especially valuable to small farmers because they can be grown on marginal lands and cultivated with minimal inputs. Cassava (*Manihot esculenta*), potatoes (*Solanum tuberosum*) and sweet potatoes (*Ipomoea batatas*) are native to Latin America, and yam (*Dioscorea* species) originates in West Africa. In some parts of sub-Saharan Africa, cassava provides over 70 per cent of the population's daily calorie intake, and in India large quantities of cassava are transformed into starch for the small-scale manufacture of sago, a low cost food for the rural and urban poor. Sweet potato, in contrast, is principally grown in China as a source of starch for noodle manufacture and in East Africa, where it is an important foodstuff. All root crops are bulky and perish relatively quickly. This means that they cannot be stored for long periods, or transported over long distances to market.

Nutritional value

Roots and tubers are valued for their high starch content (20–40 per cent fresh weight) and dietary fibre. They contain very little fat or protein and few minerals or vitamins. Some, notably cassava and certain yam species, contain varying levels of the toxin cyanide. The yellow variety of sweet potato contains appreciable quantities of ß-carotene, which is converted into vitamin A in the body. Potatoes contain moderate levels of vitamin C, and for those who consume large quantities of potatoes, this provides a significant contribution to the daily intake.

In the raw state, the starch contained within roots and tubers is indigestible, and all root crops and tubers require some form of processing to make them edible. In addition, a range of value-added products are produced from starchy roots and tubers.

Cassava

Many varieties of cassava contain potentially toxic cyanide compounds which have to be removed before consumption. Cassava processing is extremely varied (over 100 different products have been identified in six of the major producing countries in Africa), and a wide range of equipment is used, most of it very simple and locally made. Cassava is most commonly prepared into flour or dried pieces, fermented products, roasted granules and steamed granules. Other processes produce starches, beers and cassava leaf foods.

Many processing methods involve a natural fermentation, which plays an important role in the removal of toxic components and involves the production of lactic acid, which adds flavour to an otherwise bland product. Anaerobic fermentation (without air) is brought about in one of two ways: in the first, large pieces of washed, peeled cassava are left to soak under water in ponds or small containers for up to five or six days, until the root tissues ferment and soften. The second method involves peeling and grating or pounding the cassava to produce a mashed product, which is placed in closely covered sacks or other containers,

Cassava roots deteriorate rapidly, and therefore need preserving

and left to ferment for one to three days. The cassava is then further processed by dewatering, using simple wood presses to reduce the moisture content, followed by pounding and grinding to pastes which are boiled, steamed, roasted or fried, or dried to granules.

During aerobic fermentation, cassava roots are piled in a heap for several days to promote the growth of moulds and yeasts. During this period they become covered in a thick mat of fungal growth. The roots are then dried for two to fourteen days and their surfaces scraped with a knife to remove the discoloration brought about from the fungal fermentation. The sun-dried products from both anaerobic and aerobic fermentations are milled to produce flours used in the preparation of a variety of dishes. For example, *gari* is a dried, fermented product that is very popular in West Africa and eaten as a main meal with soup or stew.

Gari

Gari is a creamy white granular flour that has a slightly fermented flavour and sour taste. When properly prepared and stored it has a shelf life of six months or more. The tuber is grated and fermented to remove the cyanide compounds and then heated to drive off the cyanide gas, to destroy enzymes and micro-organisms and to dry the product.

Process notes

Raw material

Harvested roots are peeled, washed and then grated to form a pulp.

Fermentation

The grated cassava is packed into hessian or polypropylene sacks and left to ferment for two to four days at room temperature. Fermentation also results in the production of volatile compounds that give the *gari* its unique flavour. Most of the juice from the cassava pulp is expressed during this period. The time allowed for fermentation is critical: too short and the detoxification process is incomplete, resulting in a potentially toxic product; too long and the product will have a strong sour taste and the texture will be poor.

Process flow sheet for *gari*

Stage	*Equipment required*	*Directory section*
Cassava tubers		
↓		
Wash		
↓		
Peel	Peeler	29.0
↓		
Grate	Manual or motorized cassava grater	20.0
↓		
Ferment		
↓		
Press	Manual screw press	30.0
↓		
Sieve	Sieve	18.2
↓		
Roast	Open pan	27.0
↓		
Cool		
↓		
Sieve	Sieve	18.2
↓		
Pack	Heat-sealer	26.1
↓		
Store		

Pressing
The bag of fermented pulp is pressed in a manual screw press. During pressing, if too much liquid is pressed from the grated cassava, the gelatinization of the starch during roasting is affected and the product is whiter. If too little liquid is removed, the formation of granules during roasting is affected and the dough is more likely to form into lumps. The ideal moisture content is 47–50 per cent, which is assessed visually by experienced *gari* producers.

Roasting
The fermented pulp is sieved and roasted in wide shallow metal pans at 80 °C, with constant stirring for 20 to 30 minutes to prevent burning and the formation of lumps.

Roasting also adds flavour to the *gari* and gelatinizes the starch granules. If lower temperatures are used the product simply dries and produces a white powder. Too high a temperature will cause charring of the product and make it stick to the pan. After roasting, the *gari* is very hygroscopic (it has the ability to absorb moisture from the air), and should be packed in airtight, moisture-proof bags.

Sieving

Is essential to remove fibrous contaminants and produce uniform sized particles.

Quality assurance

The main points are as follows:

Raw material

Cassava roots should be thoroughly washed to remove all traces of sand, dirt and peels.

Fermentation

Fermentation should be properly controlled since too short a time will result in incomplete detoxification and a bland product. Too long a period and the product will have a strong sour taste. Both over- and under-fermentation also badly affect the texture of the *gari*.

Pressing

If too much liquid is pressed from the cassava the gelatinization of starch during roasting is affected and the product is whiter. If insufficient water is removed, the formation of granules during roasting is affected and the dough is more likely to form into lumps. The ideal moisture content is 47–50 per cent, which is visually assessed by experienced *gari* producers.

Sieving

Sieving is important to obtain a high quality product that is free of fibrous contaminants and has similar sized granules.

Roasting

The granules must be roasted to about 80 °C to achieve partial gelatinization of the starch. If lower temperatures are used, the product dries and produces a dry white powder. Too high a temperature will cause charring of the product and make it stick to the roasting pan.

Cassava starch

Freshly harvested tubers are peeled, washed and grated into a pulp. The pulp is put into a basket over a pan or bucket covered with a piece of clean cloth. Water is poured over the basket and the starch is washed out. This process is repeated until all the starch has been removed by rinsing, after which it is sun-dried. Another traditional method is to put the grated tubers in bags and to pour enough water over them to soak the contents. The bags are then squeezed and a white liquid is expressed, which is poured into buckets. More water is added to the grated cassava and the process repeated until the water runs clear. The liquid is left for several hours to allow the starch to settle, and the liquor is poured off. The remaining starch is washed three or four times until the water runs clear, and it is then sun-dried to a final moisture content of 15 per cent.

Both of these methods are simple and require very little equipment, but they are labour intensive, and productivity is low. Also, the time left to settle the starch can be as long as 24 hours, during which time fermentation may begin, resulting in spoilage and off flavours of the starch. The potential for pollution of rivers from cassava starch production is high, and there is a need to have water treatment facilities at all but the smallest scales of production.

The irregular shape and size of the roots makes small-scale mechanization of the process difficult, but there are specific pieces of machinery available. Peeled tubers are grated or rasped in cassava graters (Directory section 20.0) to break down the cell walls and release starch granules. This is an important stage in the process since the starch yield is dependent on efficient rasping. After pulping, starch milk is

Small-scale starch production

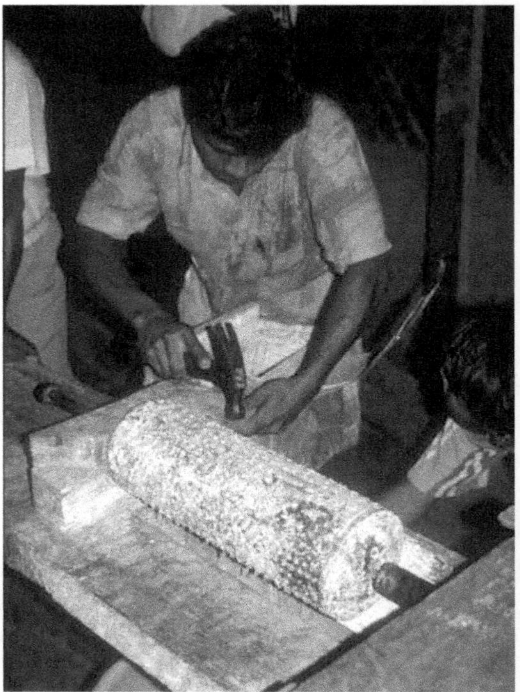

A cassava grater

separated from the fibrous residue with screens. The extracted milk is stored in large settling tanks to allow the starch to sediment out and sink to the bottom. It can be speeded up by using inclined tables or centrifugal separators (Directory section 6.0). Once it is separated out, the starch has to be dried. This is carried out in two stages: an initial de-watering stage, during which time the moisture is reduced to about 40 per cent, followed by drying to a final moisture content of 15 per cent. Various dryers (Directory section 14.0) are used, depending upon local availability, resources and scale of operation. Sun-drying has the advantage of whitening the starch through the bleaching effect of ultraviolet rays, but it also results in a more contaminated product. Dried cassava starch is pulverized into a fine powder in a hammer mill (Directory section 23.3) and then packaged into airtight bags to prevent contamination and uptake of water.

Cassava doughnuts

Cassava doughnuts are fried products which are approximately 8 cm in diameter and 3 cm thick. The texture is stiffer than cake and the crust is soft with a deep, uniform brown colour. They have a shelf life of a few days under correct storage conditions and are used as a snack food. The main quality factors are the colour and fineness of the flour, and freedom from dirt, mould and insects. Oil used for frying should be clear, of good quality and free from rancidity. Heat during frying destroys most contaminating bacteria and the soft crust restricts recontamination during storage.

The product should be cooled before packaging into a moisture-proof, oil-resistant bag to prevent contamination by soils, insects and other contaminants. It should be stored in a cool, dry place away from sunlight which would accelerate rancidity of the oil in the product.

Process flow sheet for cassava doughnuts

Stage	*Equipment required*	*Directory section*
Cassava flour		
↓		
Mix ingredients to a smooth dough	Mixer (optional) Scales	24.0 39.0
↓		
Shape		
↓		
Fry	Thermometer	38.5
↓		
Drain		
↓		
Cool		
↓		
Pack	Heat-sealer	26.1
↓		
Store		

Process notes

Mixing
Mix together 1 kg cassava flour, 75 g baking powder, 200 g sugar, 6 eggs, 125 g margarine, ½ teaspoon salt and ½ litre milk or water.

Shaping
Take mixture in spoonfuls and shape into rounds with a hole in the middle.

Frying
Deep fry the doughnuts in hot oil (approximately 150 °C) until golden brown. Remove excess oil.

Packing
Cool to room temperature. Pack in plastic bags. Store in a cool and shaded place to slow down rancidity.

Quality assurance
The main points are as follows:

Raw ingredients
Ingredients must be weighed accurately and mixed thoroughly, as even small variations can cause large differences in the final product.

Frying
Time and temperature of frying control the colour, texture, flavour and moistness of the product. Oil for frying should be of good quality and should be changed regularly. The oil affects the taste, texture and keeping quality of the final product. Any rancidity in the oil (see Part I) results in strong flavours and odours which are transferred to the fried food, thus spoiling the taste and flavour of

Dried noodles are a popular product made on the household level

the product. Fresh oil produces a higher quality product and oils should not be used more than a few times. It is important not to allow the temperature of oils to reach the 'smoke point', as this increases the rate at which they deteriorate. Suitable temperatures for frying are 180–200 °C. For greater control over the temperature of the oil and the quality of the final product, an electric thermostatically controlled deep-fat fryer can be used.

Sweet potato

Starch

Around 130 million tonnes of sweet potato are grown annually, and about 85 per cent of this is produced in China, where a large proportion of the crop is processed by villages to make starch as an ingredient of transparent noodles. Freshly harvested roots are washed and then ground, using a pin mill, hammer mill or a traditional root rasper (Directory sections 23.3, 20.0), to produce a mash. The choice of mill depends on the scale of operation and whether an alternative use for the mill exists, for example for rice or maize. After grinding, the mash is washed on screens to separate starch from the fibrous root residue. The starch slurry or 'milk' is then allowed to sediment naturally, and the residue is sun-dried or moved to a silo to be used as animal feed.

The sedimented crude starch is purified by washing and further sedimentation or by using a process known locally in Sichuan Province as the 'sour liquid method'. This involves the separate preparation of milled and fermented legume extract, which is added to the wet sedimented starch. The process decolourizes the crude starch by removing carotenes. Following the final sedimentation stage, the wet starch cake is drained overnight on a screen and then carefully broken up and sun-dried for six to eight hours under good conditions, or up to six days when the weather is poor.

Transparent noodles

This is a popular product made at the household level, by adding sweet potato starch to pre-gelatinized legume or maize starch. The 'dough' is then loaded into a saucepan type 'former' which is held above a large pot of boiling water. The dough falls through the perforations in the former into the water, where the noodles gelatinize and set. They are then washed in cold water to make them strong and transparent before being cut to size and hung on bamboo racks to sun-dry in about one day, or three to four days when the weather is poor.

Potato products

Dried potato

Papa seca is an example of a dried potato product, popular in the Andean highlands of Peru. Its production involves boiling and peeling potatoes, followed by sun-drying.

Process flow sheet for potato crisps

Stage	*Equipment required*	*Directory section*
Potatoes ↓		
Wash ↓	Vegetable cleaners	7.1
Peel ↓	Peeling equipment	29.0
Slice into thin slices ↓	Slicing equipment	10.1
Deep fry in hot oil ↓	Deep fat fryer	27.2
Drain on absorbent paper ↓		
Add flavourings (e.g. salt) ↓		
Pack ↓	Sealing equipment	26.1
Store		

Details of drying are given in Part I.

Fried potatoes

Potato crisps are a snack food made from deep-fried potato slices. They may be flavoured with salt or synthetic flavourings and packaged in polypropylene bags to provide protection against moisture uptake, air and light, which would cause rancidity. The main preservative action is heat during frying and the low moisture content of the product.

Other root vegetables and starchy fruits such as sweet potato, cassava, plantain and breadfruit can also be used to make crisps.

Process notes

Raw materials

Use potatoes that are suitable for crisp making (potatoes with a higher sugar content are most appropriate as they produce a more golden colour when fried). Clean thoroughly to remove all traces of soil and dirt.

Peeling

Abrasive potato peelers can be used. Soak or submerge peeled potatoes in clean water until ready to use to prevent blackening.

Slicing

Slice as thinly as possible. Ensure slices are

Slicing potatoes to make potato crisps

of uniform thickness so that they all cook at the same rate. Dry before adding to the hot oil.

Frying
Heat the oil to 180–200 °C (a large pan can be used, but for greater control use a thermostatically controlled fryer). Do not put in too many crisps at one time as this lowers the temperature of the oil and the crisps may all stick together. Fry until golden brown.

Draining
Remove crisps from the hot oil. Drain on absorbent paper or use a centrifugal spinner to remove excess oil. Add flavourings if required.

Packaging
Fried crisps are very hygroscopic, that is, they rapidly absorb water from the surroundings and become soft. They should be packaged as soon as they are cool (do not seal in packages before they are cool as condensation will form on the inside of the packet and make the contents soft). Packaging should be airtight and moisture proof.

Storage
Store packaged crisps in a cool place, away from direct sunlight and strong aromas. Properly packaged, they will have a shelf life of about six months.

Quality assurance
The main points are as follows.

Frying
The oil affects the taste, texture and keeping quality of the final product. Any rancidity in the oil (see Part I) results in strong flavours and odours which are transferred to the fried food, thus spoiling the taste and flavour of the product. Fresh oil produces a higher quality product and oils should not be used more than a few times. It is important not to allow the temperature of oils to reach the 'smoke point' as this increases the rate at which they deteriorate. Suitable temperatures for frying are 180–200 °C. For greater control over the temperature of the oil and the quality of the final product, an electric thermostatically controlled deep-fat fryer can be used.

Draining
Fried products should be drained to remove any excess oil. Poorly drained products are unattractive as the oil leaves a greasy film on the packaging material and accelerates rancidity and deterioration of the product. Fans and absorbent paper can be used to remove oil. Higher temperatures during frying cause less oil to be retained on the product, and new oil sticks to the product less than old oil.

Cereals and pulses

Cereal grains are the seeds of cultivated grasses, and all have a similar structure and chemical composition. Climate, soil type and tradition determine the distribution of the major food grains. Some, such as barley and oats, are mainly grown in temperate regions, while others, such as wheat and maize, are cultivated over a wide range of latitudes. Rice is the only major cereal that is consumed in the form of the whole grain, thus reducing the cost and time for processing. Sorghum and millet are prominent in arid and semi-arid regions, where they can tolerate poor soils and periods of drought, producing a yield in conditions when other crops would fail. In Latin America and many areas of sub-Saharan Africa, maize is the preferred cereal.

Together with the starchy root crops, cereals and pulses (or legumes) are the main staple crops that form the bulk of the diet in most developing countries. They are a relatively cheap source of energy, protein, vitamins and minerals. They have a low moisture content and are relatively stable during storage. Processing is primarily to transform the raw material into different products that add variety to the diet.

Although wheat flour is the most widely used type of flour for the manufacture of pastas and noodles, breads, cakes, biscuits and pastries, flour from a range of cereals, sometimes mixed with pulse grain flours and/or flours from tuber crops can also be used.

Nutritional value

The whole grains of cereals and pulses provide energy in the form of carbohydrates and

Cereals and pulses are processed into a range of products

fats, together with a variable amount of protein depending upon the species and cultivation conditions. Much of the carbohydrate in pulses is in the form of complex sugars which are not readily digested and can cause problems with indigestion and flatulence. Most pulses contain 20–40 per cent protein, compared to cereals which contain 7–15 per cent. Cereal protein has good nutritional value, although inferior to animal protein because some amino acids are missing (see below). This can be corrected by combining cereals with pulses, and it is no accident that most traditional food systems are based on a combination of cereals and pulses (e.g. rice and dhal in Asia, maize and beans in Latin America and Africa, rice and peas in the Caribbean and cassava and beans in Africa). Cereals are a good source of the B group of vitamins, but maize is a poor source of niacin, and pellagra (a disease associated with niacin deficiency) is common in communities that subsist on maize.

Amino acids and proteins

There are many different types of protein and all are made up of long chains of amino acids. Some of these amino acids are found at low levels in plant foods, in particular two called lysine and methionine. Cereal proteins are low in lysine, whereas legume proteins are low in methionine. By mixing cereals and legumes the amino acids in the mixture are nutritionally balanced.

Many legumes contain anti-nutritional factors, which in red kidney beans and lupin seeds are poisonous if consumed. Trypsin inhibitors are not themselves poisonous, but they prevent the action of trypsin (one of the digestive enzymes) and therefore prevent the body from absorbing amino acids from the food. They are present in soybeans and to a lesser extent in cowpeas and winged beans. Trypsin inhibitors are not heat stable and can therefore be removed by soaking and heating. Other pulses, such as lima bean, contain cyanide, but they can be made safe provided that they are soaked and cooked before consumption. Tannins in legumes reduce the digestive capacity by combining with either protein in the diet or with digestive enzymes. Generally, tannins are located within the seed coat and are more concentrated in seeds with dark coloured skins. They can be removed or reduced by soaking or by removing the seed coat. Tannins are also implicated in the 'hard-to-cook' problem that occurs in some types of legumes. These legumes require substantial cooking time to soften and become edible, which is expensive in terms of the time taken and the fuel consumed, and has significant implications for the cost of processing, and at a domestic scale on fuelwood consumption. Fermentation reduces the preparation time and the amount of fuel consumed, while at the same time improving the nutritional value and adding variety to the diet. Another method is to roast the beans and then mill them to a flour, which removes anti-nutritional factors and speeds up the cooking process, so reducing energy consumption. It is an ideal process for making combined cereal/pulse weaning foods.

Weaning foods

Traditional weaning foods around the world are based on porridges made from the local staple (e.g. maize, banana, rice, plantain, cassava). These have a low nutritional value and are prepared under non-hygienic conditions. The requirements for a good weaning food are as follows.

Nutritional requirements:

- high energy content
- low viscosity (i.e. an acceptable thickness/consistency for young children)
- balanced protein (containing all essential amino acids)
- adequate levels of vitamins and minerals
- no anti-nutritional components.

Physical requirements:

- easy and quick to prepare
- easy to consume and a pleasant taste
- adequate shelf life
- made from local ingredients
- affordable
- safe microbiological quality.

Weaning foods can be prepared from a combination of cereals and pulses by roasting and milling the mixture and packing as a dry mix. This can then be reconstituted with water and mixed to a porridge-like consistency. However, it is essential that nutritional advice is sought to ensure that the mixture of cereals and pulses is correctly balanced to contain the correct proportions of amino acids and sufficient energy, protein, vitamins and minerals to maintain the health and growth of the child. Hygiene is of utmost importance during the preparation of weaning foods to prevent diarrhoeal disease and mortality among infants.

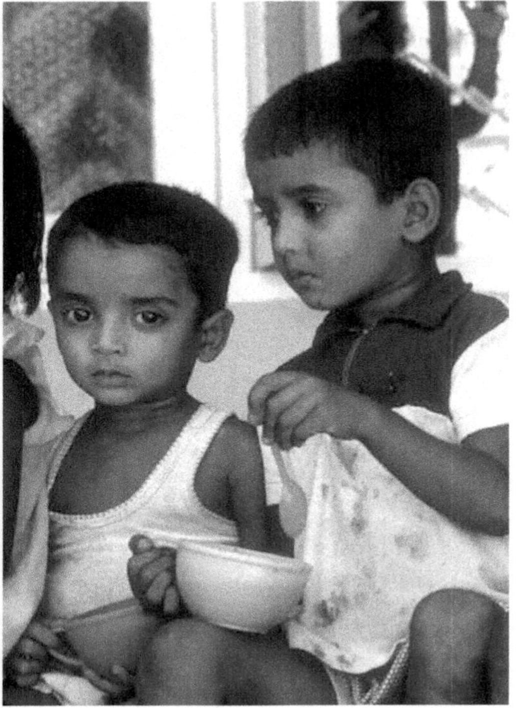

Weaning is an important stage of an infant's life

Weaning foods can also be fortified by the addition of a little oil to a traditional starchy porridge. This increases the energy value and improves the flavour and palatability of a bland food. If red palm oil is used, there are additional benefits from the high β-carotene content. Green leaves are another rich source of β-carotene and also contain iron. They can be dried and ground and added to traditional cereal-pulse based weaning foods to increase their nutritional value.

In Ghana *gratimix*, a high protein weaning food made from groundnuts, maize and beans, has been developed for small-scale production. The three components are cleaned separately and roasted in an open pan. The groundnuts are de-hulled and the ingredients are mixed in the following proportions: 70 per cent maize, 15 per cent cowpea and 15 per cent groundnut, and then milled using a corn mill (Directory section 23.0) to produce a fine flour. The flour is packed into double thickness polythene sachets and heat-sealed to give a shelf life of three months. Weaning biscuits can be made from pulses (beans), oil-seed (groundnut or sesame) and cereal (maize, millet or wheat flour). The oil-seed and cereal are separately roasted and ground to a fine paste or flour, and beans are soaked to remove the skins and ground to a flour. One part each of the oil-seed paste, bean flour and cereal flour are mixed together. A half part of sugar, half part of oil and 1 teaspoon of baking powder are added for every 10 spoonfuls of mixture to form a dough, using a powered mixer (Directory section 24.0). This is then formed into biscuits and baked in an oven (Directory section 25.0) for 15 minutes.

Sprouted grains, such as barley, millet or sorghum, are rich in an enzyme called amylase, which breaks down starch into sugars. These can be added to a starchy porridge to produce a sweeter, more palatable food with a thinner consistency that is easier to digest. Cereal grains are allowed to sprout and are then dried and ground into a powder. A small pinch of the powder is mixed into the starchy porridge and after a short time the viscosity of the porridge decreases to become more liquid. The only downside is that sprouted grains are traditionally used in the malting process to produce beer, and for this reason some cultures find them unsuitable for the preparation of weaning foods. However, there is no alcohol produced during sprouting, only the conversion of starch to sugars.

Process flow sheet for primary processing of cereals and pulses

Stage	*Equipment required*	*Directory section*
Harvesting ↓		
Pre-drying in the field ↓		
Threshing ↓	Threshers	40.0
Winnowing ↓	Winnowers	42.0
Drying ↓	Crop dryers	14.0
Storage ↓	Storage structures	
Cleaning ↓	Seed cleaners	7.2
Grading ↓		
Hulling ↓	De-hullers	12.0
Pounding, milling ↓	Plate mills Hammer mills Roller mills	23.1 23.3 23.2
Tempering/soaking ↓		
Parboiling (rice) ↓		
Drying ↓	Dryers	14.0
Sieving ↓	Sieves	18.2
Storage		

Processing

Further details of the technologies and equipment used for primary processing are given in the companion volume *Tools for Agriculture* (Carruthers and Rodriguez, 1992) and the ITDG/UNIFEM publication *Cereal Processing* (UNIFEM, 1994).

Milling

Milling is included here as a common small processing operation in developing countries. Grains and pulses are milled to produce flours which are used to produce many types of food. Three main types of mill are available:

- plate mill
- roller mill
- hammer mill.

The choice of mill depends on the raw material and the scale of production. Hammer mills are almost universally used throughout the developing world, and in West Africa plate mills are widely available. Roller mills (for wheat milling) are not used at the small scale because of their high cost and maintenance requirements. The equipment directory (Directory section 23.0) illustrates a range of different mills. Details of the milling process and guidelines on choice of machinery are given in Carruthers and Rodriguez (1992).

Cereal and pulse flours and flakes are the starting materials for the following types of secondary processing: fermenting, baking, puffing, flaking, frying and extruding.

Fermentation

Fermented cereal doughs are traditional products in many countries. For example, fermented steamed maize dough such as *kenkey* in Ghana, *bagone* in Botswana, *banku* and *ugali* in Western and Eastern Africa respectively are each staple foodstuffs. Fermented legume products are particularly important in regions where little or no protein of animal origin is consumed, either for economic or religious reasons. Two of the most important fermented legume products are soybean curd (or tofu) and Indonesian *tempeh*. Both are prepared from

Traditional method of pounding grain

soybeans and are highly nutritious. Fermentation also causes an increase in acidity which improves the keeping qualities and increases the safety of foods by retarding the growth of pathogenic micro-organisms (see Part I).

Tempeh is prepared by fermentation of cooked de-hulled soybeans by the mould *Rhizopus oligosporus* at 35 °C for 48 hours. This makes the protein more digestible and reduces the amount of trypsin inhibitors. Other benefits include the production of vitamin B12 and anti-diarrhoea factors. The product has a shelf life of 3–4 days at room temperature, and is normally fried as a vegetable or meat alternative, dried and used in soups and as a condiment, or cut into thin wafers and deep fried in oil and served as a biscuit. Other pulses such as Australian sweet lupin and chick pea, cowpea and pigeon pea can also be used.

Fermented wheat doughs are made from a combination of flour, fat, yeast, and water/milk. By including other ingredients such as sugar, fruit and nuts there is scope to create a wide variety of products. The production of doughs involves mixing ingredients together to a smooth and uniform consistency. This can be done manually, or with a powered mixer. The type of mixer depends upon the product being prepared, and there are a large number of different designs (Directory section 24.0). There are various ways in which fermented doughs may be subsequently processed. Batters are produced from similar ingredients and have a more liquid consistency. They are not often fermented, but are otherwise processed in a similar way to doughs.

Toasting and baking

The action of heat on starchy foods causes the starch to pre-gelatinize, which makes it easier to digest. Pre-gelatinized starch also takes less time to prepare. For example, porridge made from toasted maize will cook in 10 minutes compared to 1.5 hours for untoasted maize.

A wide assortment of baked products can be produced from fermented doughs, including hundreds of different types of bread, doughnuts and pizzas. Unfermented doughs and batters are used to produce cakes, pancakes, biscuits and unleavened breads (e.g. chapattis, *roti*, and tortillas), which are baked using a hotplate (or 'griddle'). Once baked, foods such as breads and pastries have a shelf life of 2–5 days, whereas others, such as biscuits and some types of cake, have a shelf life of several months if correctly packaged. The purpose of baking is therefore not for preservation but to change the eating quality of staple foodstuffs and to add variety to the diet.

Doughs are formed to the required shape and the fermentation is allowed to take place at a constant temperature in a proofing oven (Directory section 25.0) before baking. A wide variety of designs of baking oven exist (Directory section 25.0). Batters are either dropped onto the griddle (e.g. pancakes and *kisera*, a Sudanese product made from sorghum flour) or contained in baking tins (e.g. cakes). During baking, food is heated by the hot air in the oven and heat from the floor and walls of the chamber. Moisture at the surface of the food is evaporated by the hot air and a dry crust is produced. In biscuit production more moisture escapes from the interior of the food before it is sealed by the crust, leaving a dryer product. The charts below show details of the production of an unfermented product (tortillas) and a fermented bread.

Tortillas

Ingredients: 10 kg whole clean common maize and 70 g lime powder (calcium hydroxide).

Process notes

Raw materials

The grain used should be of good quality. High-moisture or mouldy grains should not be used. Yellow and white maize can be used, depending upon local availability. The lime powder (calcium hydroxide) should be dried and free from dirt or contamination. Tortillas can also be prepared from instant tortilla flour by adding water to make the dough.

Boiling

Maize is boiled in water with lime for one hour. Ensure that the water covers the grain at all times. The addition of lime to the boiling water helps to soften the grain.

Washing

Boiled grains are washed in running water to remove the hulls and excess lime. Washed grains are drained before being milled.

Milling

The grains are wet-milled using a plate mill to form a soft paste or dough.

Divide and shape

The dough is divided and shaped, either by hand or using a tortilla press, to make a

Process flow sheet for tortillas

Stage	*Equipment required*	*Directory section*
Maize		
↓		
Boil with lime (calcium hydroxide)	Boiling pan	27.0
↓		
Wash		
↓		
Mill with the addition of water	Plate mill	23.1
↓		
Mix to a dough	Mixer	24.0
↓		
Shape		
↓		
Bake on a hotplate		
↓		
Pack	Heat-sealer	26.1
↓		
Store		

round pancake about 10 cm diameter. The thickness of the pancake varies according to local taste.

Heat
The tortilla is cooked by griddling on both sides on a clay hotplate until it is golden in colour (about 1 minute).

Cool and pack
Cool to room temperature and pack in quantities of 6 or 12 in polythene bags.

Store
Store in a cool, dry, well ventilated place. Tortillas are generally eaten fresh when produced and have a shelf life of only one or two days.

Quality assurance

The main points are as follows.

Raw materials
Grains should be adequately hulled and milled to produce a consistent and smooth dough.

Shaping
The dough should be formed into uniform pieces.

Heating
The time and temperature of griddling affects the colour and texture of the product.

Cooling
Tortillas should be cooled before packaging to prevent the build up of condensation inside the package.

Hygiene
The mill and all equipment used should be well cleaned. If the tortillas are formed by hand, good hygienic practices are necessary to avoid contamination by bacteria.

Injera

This is a traditional Ethiopian fermented bread made from tef (*Eragrostis tef*), or other cereals such as sorghum, millet, barley and wheat. It is a soft, thin, flat disc measuring 60 cm in diameter. The top has uniformly spaced honeycomb-like 'eyes' and the underside is smooth. It can have a variety of colours from whitish-cream to reddish-brown, depending on the type of cereal used. It tastes slightly sour and is served along with *wet*, a traditional Ethiopian stew.

Tef is winnowed, sieved and ground in a stone mill (Directory section 23.0). The flour is mixed with water and a starter culture (fermented dough saved from a previous batch), covered and left to ferment for three days. A small amount of the fermented dough is added to water, mixed and boiled to produce *absit*. The main dough is thinned by adding an equivalent amount of water to the original weight of flour and the *absit* is added to produce the final batter. The batter undergoes a short second fermentation for about 30 minutes before baking on a *metad* (clay griddle – Directory section 25.0) for 2–3 minutes. *Injera* can be stored in layers in a straw basket with a tight cover for up to three days.

Bread

The ingredients required for leavened bread are wheat flour, salt, oil, sugar, water and

Injera *is a flat bread made from fermented tef grain*

yeast. Other ingredients such as milk, cheese, fried onions, dried tomatoes, herbs, cinnamon or dried fruits may be added to change the flavour, appearance or keeping qualities of the bread. Wheat flour is used because it contains proteins which form a strong, elastic network in the dough. As the yeast produces carbon dioxide gas, the protein network traps it inside the dough, to give the typical aerated texture of raised bread. Salt is added to give flavour and to help the elastic dough to form.

Typical ingredients for a standard white bread are as follows (quantities given are for a commercial batch):

strong flour	90 kg
dried yeast	910 g
fine salt	1.59 kg
granulated sugar	850 g
baker's fat/margarine	910 g
water	53 kg

Production schedules for other baked goods can be found in Bathie (2000).

Process notes

Mixing

Disperse the yeast and a little sugar in a portion of water (approximately 5–10 times the weight of yeast) at 35 °C. Stir vigorously and leave for 12–15 minutes to allow the yeast to activate.

Add the salt and remaining sugar to the dough water and add to flour mixture. Add the activated yeast and continue mixing until an elastic dough is formed.

Proving

Cover with a damp cloth to prevent the dough drying out and hardening. Place it in a warm cabinet (about 26–30 °C) for 45 minutes to allow the yeast to ferment and the dough to rise.

Knock back

Knead the dough to expel the gas.

Bread is suitable for small-scale production

Process flow sheet for bread

Stage	*Equipment required*	*Directory section*
Ingredients		
↓ Sieve flour and mix in fat	Sieve	18.2
↓ Mix in salt, water, sugar	Dough mixer	4.1, 24.0
↓ Mix in activated yeast		
↓ Proof	Proofing cabinet	4.0
↓ Knock back	Bread forming tables	4.0
	Bread moulds	4.2
↓ Form	Dough divider	4.2
	Moulding machine	4.2
	Baking trays	4.0
	Scales	39.0
↓ Final proof	Proofing cabinet	4.0
↓ Bake	Oven	25.0, 25.2
↓ Cool	Cooling racks	4.0
↓ Pack	Heat-sealer	26.1

Form
After a further 15 minutes divide the dough into equally weighed pieces and mould into balls either manually or with a mechanical dough dividing machine and moulder, and place in greased baking tins.

Proving
Cover with a damp cloth and rest for 10 minutes in a proofing cabinet for final proofing.

Bake
Bake at 220 °C for 25–30 minutes for a 500 g loaf or 40 minutes for an 800 g or 1 kg loaf.

The bread may be brushed with milk to give a shiny glaze to the top. The baking time depends on the size and type of bread, the number of loaves in the oven and the type of oven. When properly baked the loaf should make a hollow sound when tapped on the base with the fingers.

Cool
Remove the bread from the oven, de-pan, set on cooling racks and allow to cool.

Pack
Pack in paper or polythene bags depending on length of storage time before sale or consumption.

Quality assurance
The main points are as follows.

Raw ingredients
The flour must be free from foreign matter, soils and other contaminants such as weevils. Good quality active yeast must be used in order to ferment the dough to the desired level within the required time. Flour, water and yeast should be in the correct proportions to produce the required texture.

Weighing
Accurate weighing of ingredients is an important control point because small variations can cause significant differences in the final product.

Mixing
Thorough mixing and kneading of the dough are required to produce a uniform product.

Baking
Temperature and time of baking are important to ensure the correct colour, texture and flavour of the product.

Cooling
The product should be properly cooled before packing because condensation of moisture on the inside surface of the pack encourages mould contamination. Polythene or paper bags provide protection against some insects and soil contamination. The product should be stored in a ventilated, cool, dry place, raised off the floor.

Thorough mixing and kneading of dough

Small-scale bakers do not normally have facilities for flour analysis and must rely on information supplied by the miller. However, useful information can be obtained from the following simple tests, which are described in more detail by Fellows et al. (1995).

Gluten measurement
It is important that a baker buys the correct type of flour, which for bread-making should be 'strong' (i.e. made from hard wheat having a medium–high protein content). For other products 'weak' (soft wheat) flours that are lower in gluten are used. Gluten can be assessed by simply washing out the starch

from a small ball of dough under running water to leave the gluten proteins. The quality can be assessed by pulling the gluten piece apart and noting how much it stretches and its breaking point. This is a quick test, which with experience provides much useful information to the baker.

Starch gelatinization

The ability of starch to gelatinize (to thicken when heated with water) is important especially when preparing batters for cakes, as this will determine the structure and volume of the cake. This simple method can be routinely used in a bakery.

1. Mix 100 g of flour with 900 g of hot water in a pot.
2. Heat the pot until the flour mixture has gelatinized.
3. Pour into a 1 litre measuring vessel such as a graduated glass cylinder.
4. Stand the measuring vessel in hot water.
5. Drop a small steel ball into the mixture and record the time to drop through 200 ml.
6. Compare the time (in seconds) against the standard batch.

It should be noted that this method relies on comparison of results between different batches using the same measuring cylinder. It cannot be used to compare results from different cylinders unless they are identical.

Water absorption measurement (flour)

The amount of water absorbed by flour is one of the most important factors affecting the structure and texture of all baked goods, including breads, biscuits and cakes. In this method a burette is required to accurately measure added water. This equipment can usually be obtained from a scientific supplier in the capital city, but could also be made locally by carefully calibrating a glass tube to indicate exactly the volume of water contained.

1. Weigh 100 g of flour into a small mixer or bowl for mixing by hand.
2. Slowly add water from a burette until a standard dough is made. This is judged by the processor by adding water until the consistency feels right.
3. Record the amount of water added.

New batches of flour are tested alongside the existing material. This can be used as a comparative test to indicate a wrong grade of flour.

Puffing

Puffed grains and pulses are used as ingredients in breakfast foods or as snack foods. During puffing, grains are exposed to a very high steam pressure in a sealed steel container known as a 'puffing gun'. A door on the container is opened which rapidly releases the pressure and causes the grains to burst open. Operators must take special care when handling the equipment necessary for puffing to prevent accidents. Alternatively moist grains can be puffed by mixing them with very hot sand and then sieving out the sand using a metal sieve. Puffed grains and pulses may be further processed by toasting, coating, or mixing with other ingredients.

Pushto is a traditional Indian weaning food made from puffed wheat, similar to the process for puffed rice and for toasting peanuts. Cleaned, de-hulled wheat is moistened with a little water and then added to a pan containing hot sand at 250–260 °C (1 kg of grain to 10 kg of sand). The grain and sand are stirred continuously for about 1 minute until all the grain has puffed. The mix is then quickly sieved so that the grain does not burn. The puffed grains are then ground to a flour and mixed with soya flour which has also been roasted over a fire for about 6 minutes, to remove the trypsin inhibitors. Sugar is also ground to a fine powder and added to the flour, and ground dried leaves could also be added as a source of vitamin A and iron.

Flaking

Before flaking, grains are partially cooked to soften them. They are then pressed or rolled into flakes which are subsequently dried.

Such partially cooked cereals may be used as quick-cooking or ready-to-eat foods. Both flaked and puffed grains are crisp with a moisture content below 7 per cent, and require packaging in moisture-proof containers.

Frying

Frying involves cooking food in hot oil. Doughnuts are an example of a fried dough which is consumed as a snack food (see also section on Cassava). It is important to make sure that the temperature of the oil is correct. If the temperature is too low then the level of fat absorption in the doughnut will lead to a greasy product. If the temperature is too high, it is necessary to remove the doughnuts before they are cooked in the centre, to prevent over-browning. The shelf life of fried foods is mostly determined by the moisture content after frying. Doughnuts have a relatively short shelf life owing to the high moisture content, and packaging is not necessary, except to keep the product clean.

Extrusion

Extrusion involves forcing food through a hole (or die) to produce strands or other shapes such as whorls, rods, hoops etc. Once the food has been extruded it may undergo a series of further processes such as frying, boiling or drying. There are two types of extruder: a high temperature/pressure extruder that produces expanded snack foods or soya infant foods, and a low temperature/pressure extruder that produces noodles, pasta etc. (Directory section 16.0). High temperature/pressure extruders are likely be too expensive for most small-scale processors, but low temperature/pressure extruders are affordable. Examples of extruded doughs include snack foods such as 'Bombay mix' and a wide variety of pasta products and noodles. The latter are increasing in popularity because they are quick and easy to cook, offer convenience and require less fuel for cooking. A further advantage of these products is that when dried, they have a long shelf life. The table below outlines stages in production for rice noodles. Similar processes are used for pasta and snack foods using appropriate ingredients.

Dough is extruded through small holes into hot oil to make channa chur

Preparation of noodles

Preparation of noodles

Noodles can be made from a range of flours including rice, wheat, maize and potato. The dough can be processed in one of two ways, either rolled out into thin sheets of dough and cut into strands, or extruded. The strands are then steamed, and may either be eaten fresh, or processed further by drying. Drying can be achieved using either a solar or a fuel-fired dryer. It is possible to dry the noodles in the sun, but the quality of the finished product is likely to be lower. Packaging requirements for dried noodles include a moisture-proof package (e.g. polythene) and an outer carton/box to prevent crushing.

Process notes

Raw material

Rice grains should be sorted to remove broken or mouldy pieces and should be free from soils or contamination. Rice which has been put through a polisher to remove its outer layers should be used. It is milled to a fine flour using a manual or powered grinding machine.

Mixing

Gradually mix in just enough hot water to form a stiff porridge/dough.

Extrude

Dough is extruded using a hand-operated extruder with a 2–3 mm diameter die.

Dry

The long extruded noodles are dried in the shade for about six hours.

Cut

Select the dried noodles for size and colour and cut into pieces up to 15 cm in length.

Pack

Package into dry, moisture-proof bags and store in a cool dry place.

Quality assurance

The main points are as follows.

Raw material

The smoothness of the noodle dough

Process flow sheet for rice noodles

Stage	*Equipment required*	*Directory section*
Mix ingredients (rice flour and water) to a dough	Powered mixer (optional)	24.0
↓		
Roll into thin sheets of dough		
↓		
Cut into strands	Cutting equipment	10.4
↓		
Extrude	Extruder	16.0
↓		
Dry	Air dry	
↓		
Cut	Cutting equipment	10.4
↓		
Pack	Sealing machine	26.1

depends on the fineness of the flour. Therefore flour that has been passed through a 0.5 mm sieve should be used.

Mixing
Hot water is added slowly and stirred gently to make a smooth stiff dough.

Drying
Noodles are dried in the shade to avoid rapid removal of water, which could cause cracking and splitting of the noodles. When fully dried, the noodles will break when bent.

Packaging
Packaging material needs to provide a barrier to the entry of moisture. Dried noodles are very hygroscopic and will spoil (soften, turn mouldy and develop rancidity) if they take up water.

Hygiene
Potable water and normal hygienic food handling practices should be used.

Oil-seeds and nuts

The majority of nuts are dried and eaten whole, or used as ingredients in other processed foods (e.g. baked goods, confectionery products or breakfast cereals). The main type of oil-seed processing is extraction of cooking oils, and for nuts, extraction of cooking oils and essential oils is the most important, although coconut meat has a variety of other uses.

Cooking oils

There is a universal demand for vegetable oil for domestic cooking, as an ingredient for other food production (in baked goods and fried snack foods) and as a raw material for the manufacture of soap, body/hair oils, and detergents (see Table 8).

If properly stored, vegetable oil has a shelf life ranging from 6 to 12 months. Heat applied during processing destroys enzymes in the raw materials and also any contaminating micro-organisms which would cause rancidity. Additionally, the oil may be heated after extraction to remove as much water as possible. This lessens the occurrence of microbial spoilage during storage. Correct packaging and storage conditions slow down chemical changes caused by light and heat which may lead to rancidity (see Part I).

Oil-seeds and nuts have a long shelf life if kept dry, and oil processing can therefore continue throughout the year, which makes better use of the equipment than more seasonal products. The low volume of oil, compared to oil-seeds, makes transport and

Table 8. Major oils and their uses

Raw material	*Oil content (%)*	*Uses*
Oil-seeds		
Castor	35–55	Paints, lubricants
Cotton	15–25	Cooking oil, soap making
Linseed	35–44	Paints, varnishes
Niger	38–50	Cooking oil, soap making, paint
Rape/mustard	40–45	Cooking oil
Sesame	35–50	Cooking oil, tahini
Sunflower	25–40	Cooking oil, soap making
Nuts		
Coconuts	64 (dried copra) 35 (fresh nut)	Cooking oil, body/hair cream, soap making
Groundnuts	38–50	Cooking oil, soap making
Palm kernel nuts	46–57	Cooking oil, body/hair cream, soap making
Shea nuts	34–44	Cooking oil, soap making
Mesocarp		
Oil palm	56	Cooking oil, soap making

Peanuts used for oil extraction

distribution easier, and the wider area creates larger potential markets. Oil extraction produces a by-product ('oilcake') which is nutritious and can be used either for animal feed or, in the case of groundnuts, as an ingredient in making other food products.

There are disadvantages to oil extraction: the higher value of oil-seeds compared to other crops means a higher financial risk from losses; the equipment can be expensive; and year-round production needs a large working capital to buy and store seasonal crops. There is also the risk of competition from large-scale producers who are able to market high-quality oils at a lower cost because of economies of scale. Small-scale extraction produces crude oil which has a different appearance and flavour to commercially refined oils. It is therefore necessary to test market the oil for acceptability before embarking on a processing enterprise.

Not all oil-bearing seeds, nuts and fruits contain edible oil. Some contain poisons or unpleasant flavours, and these are only used for paints. Others such as castor oil need very careful processing in order to make them safe. Such oils are not suitable for small-scale processing. Oils from other crops, such as maize, are extracted by using solvents which dissolve the oil. This method of extraction is also not suitable for small-scale operation because of the high capital costs of equipment, the need for solvents which may not be easily available, and the risk of fire or explosions.

Aflatoxins are a poisonous group of compounds produced by certain moulds which grow on seeds and nuts. They are poisonous to both humans and animals (causing liver damage, cancer and death) if consumed over a prolonged period. Contaminated seeds and nuts can be easily identified if there are visible signs of mould growth, discoloration or a shrivelled appearance. Aflatoxins are not destroyed by heating or removed during subsequent processing, but the mould growth can be prevented by drying the crop correctly.

Table 9. Various stages in oil-seed processing

Processing stage	*Oil-seeds*	*Groundnut*	*Coconut (wet method)*	*Palm kernel*	*Palm fruit*	*Equipment*	*Directory reference*
Decorticate /de-husk	✓	✓				Decorticators	12.0
						Winnowers	42.0
Crack			✓	✓		Hammer mills	23.3
						Kernel cutters	10.6
Grind/grate		✓	✓	✓		Hammer mills	23.3
						Graters	20.0
						Roller mills	23.2
Pulp					✓	Pulpers	31.0
Heat/condition	✓	✓		✓		Seed scorcher	–
						Heat sources	25.0
						Thermometer	38.5
						Measuring and weighing equipment	39.0
Press/expel	✓	✓	✓	✓	✓	Ghanis	15.0
						Oil presses	30.2
						Expellers	15.0
Clarify	✓	✓	✓	✓	✓	Filters	18.0
						Clarifiers	
Pack	✓	✓	✓	✓	✓	Filling and sealing equipment	17.0 26.1

Nutritional significance

Oil provides twice as much energy as the same quantity of carbohydrate and contains a range of fat-soluble vitamins (A, D, E and K) and essential fatty acids, all of which are necessary for the healthy functioning of the body.

Processing

Oil is contained in plant cells, and is released when cells are ruptured. There are four main stages in the extraction of oil:

- preparation of the raw material
- extraction
- clarification
- packaging and storage.

Table 9 outlines the stages and equipment needed to process oils.

Process notes

Raw material preparation

Decortication is the removal of fibrous husks or seed coats prior to processing, and manual or powered decorticators are available. Winnowing takes place after decortication, and is the separation of the husks or seed coat from the oil-bearing material. Traditionally, this is achieved by throwing the seeds into the air and letting the husks blow away, but for higher rates of production winnowing machines (manual or powered) are available. Palm kernels and coconuts are cracked and the shell removed either manually using a hammer or heavy knife, or by motorized hammer mills. Groundnuts are ground to a coarse flour by pounding using a pestle and mortar or using roller mills.

Flakers are used to break sunflower seeds, and hammer mills are used to break palm kernels. Coconut flesh is grated using manual or powered graters.

Heating/conditioning
Wet raw materials, such as coconut or palm fruit, are heated to break the oil/water emulsion and allow the oil to be separated. Other raw materials, such as groundnuts and sunflower seeds, are 'conditioned' by heating with a small amount of water before oil extraction. This assists in rupturing the oil-bearing cells and decreases the viscosity of the oil, allowing it to flow more easily. The required moisture content and temperature varies according to the raw material. Groundnut flour for example needs 10 per cent added water and is heated to 90 °C in a seed scorcher. Heating is complete when the mixture stops sticking together and forms a free-flowing flour.

Oil extraction
The three main types of equipment for oil extraction are:

- motorized or animal-powered ghanis
- oil presses
- oil expellers.

A ghani consists of a wooden mortar fixed to the ground and a rotating pestle. The raw material is crushed by a rubbing action and pressure. Oil is pressed out and runs through a hole at the bottom of the mortar.

There are many types of oil press, but all work on a similar principle. Raw materials are placed in a heavy perforated or slotted metal cage and a metal plunger is used to press out the oil. The main difference in design is the method used to move the plunger, which can be either a screw thread or a hydraulic lorry jack. It is important that the mineral oil used in hydraulic presses does not contaminate the oil, and the jack is therefore better placed below the press rather than above it.

Oil expellers have a horizontally rotating screw which feeds raw material into a barrel-shaped outer casing with perforated walls. The expeller grinds, crushes and presses out the oil as the oil-seed passes through the

A ghani in operation

An oil press – the UNATA press for groundnuts

machine. Oil flows through the perforations in the casing and is collected underneath. The residue, or oilcake, is pushed out of the end of the unit. The screw requires repair or replacement at frequent intervals because of wear from the seeds, and especially if seeds are inadequately cleaned and contain sand.

Clarification
Crude oil contains fine pulp and fibre from the plant material, and small quantities of water, resins, colours and bacteria which makes it dark in colour. These contaminants are removed by clarifying the oil, either by allowing it to stand for a few days and then removing the upper layer, or by using a clarifier. If further clarification is needed, the oil may be filtered through a fine filter cloth. Finally, the oil is heated to boil off traces of water and destroy bacteria. When these impurities are removed, the shelf life of the oil can be extended from a few days to several months, provided it is stored properly.

Packaging and storage
Rancidity causes oils to develop off flavours

An oil expeller (Tinytech model)

during storage and is prevented by using clean, dry containers which exclude light and heat, and prevent contact with metals such as iron or copper (see Part I). Sealed coloured glass or plastic bottles kept in a dark box are adequate. Metal cans may be used provided the metal is tin-coated. Alternatively, glazed ceramic pots sealed with a cork and a wax stopper are also suitable. Reused bottles must be well cleaned to ensure that there is no trace of old oil on the inside, and containers should be properly dried after cleaning to remove all traces of water. If oil is adequately packaged and kept away from heat and sunlight, the shelf life can be 6–12 months.

Essential oil distillation

Essential oils for use in perfumery and flavourings can be extracted from various parts of plants including roots, leaves, stems, seeds, nuts and flowers. Many of these compounds are evaporated at the temperature of steam and can be extracted by steam distillation. Others are known as 'fixed' oleoresins, and these are extracted using organic solvents. The preparation of the raw material for distillation varies from plant to plant. Some, in particular flowers, should be distilled immediately after harvest; others have to be stored or dried and some require fermentation. Woody materials and roots require cutting and grinding to aid the release of the essential oils.

Processing

There are three types of distillation:

- water or 'hydro-distillation', where steam is generated by boiling water in a still
- water and steam, or 'wet-steam distillation'
- dry steam, in which steam is generated by an external boiler.

Water distillation is the simplest method and is widely used in developing countries. The material is placed in the still (Directory section 43.0), covered with water and boiled. The essential oils leave the still via a condenser.

Distillation of cinnamon oil

Open fire heating can cause problems due to lack of control and burning which result in poor quality oils.

In water and steam distillation, the raw material is supported on a mesh above water which boils in the base of the still. Wet steam passes through the material carrying away the essential oil. If open fire heating is used, there is little chance of burning the material. It is important that steam is able to pass through the material, and several screens may be used to support the material and stop it packing down.

Steam distillation is the most advanced method and has advantages in greater fuel efficiency and controllability, and high quality oils. Steam is generated in a boiler and passed to a perforated coil in the base of the still. The charge is held on screens, as in water–steam stills.

In each method the steam carrying the oil leaves the still body via a pipe called a 'gooseneck'. A mesh is frequently placed in the gooseneck to avoid material being blown out of the still by the steam pressure. After leaving the still, the steam is condensed to water by cooling it in a condenser. Two types of condenser are used: coil condensers are easy to make and low cost. However, they use a considerable amount of water, are not efficient and can generate back pressures to the still. Tube condensers are efficient, have low back pressure and use less cooling water, but they are expensive and need a good workshop for fabrication. The metal used for constructing a condenser depends on the type of oil being distilled because it must not react with the oil. Stainless steel is the ideal material, but copper or tin plated copper is acceptable. Mild steel is rarely suitable.

When the steam condenses, the essential oil separates and may float to the surface or sink. A special separator, called a Florentine flask, is used to trap the oil while allowing the water to drain away. There are two types of Florentine flask: one for light oils which float on the water and another for heavy oils which sink. Some oils, such as bay leaf oil, have both light and heavy components and require great skill from the operator to change the Florentines as the distillation proceeds. Several Florentines may have to be used to completely recover the oil.

The water temperature at this stage can be critical to high recoveries and in many primitive stills the temperature of the condensate is too high for good yields. High temperatures can also lead to the loss of the lightest part of the oil which may have an important aroma characteristic. At the end of the distillation the oil is separated in a glass separating funnel, dried by filtering it through cotton wool and packed in sealed glass bottles.

Other products

Nata de coco is a gelatinous material derived from coconut milk, which is extracted from ground coconut meat. It is used for the production of sweets for desserts. After the milk is extracted the white residue can be made into several types of confectionery or snack foods, including macaroons, coconut brittle and cocoballs (see also the section on *Honey, syrups and sugar confectionery*).

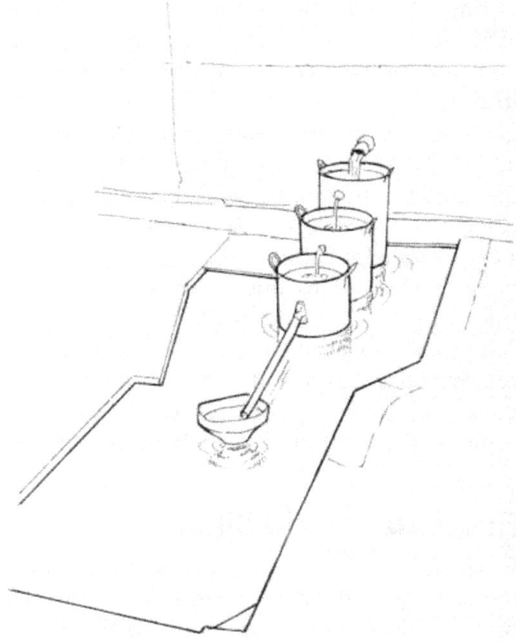

Florentine flasks used for oil distillation

Coconuts are split and the coconut water is used for making vinegar or for drinking. The coconut meat is removed from the shell using a manual or electric grater (Directory section 20.0). Grated coconut meat is wrapped in cheesecloth and squeezed manually to extract the milk. A batch of *nata de coco* is formulated by adding the following ingredients to each kg of milk:

refined sugar	2 kg
glacial acetic acid	400 ml
'mother liquor'	5 litres (contains the micro-organism that will initiate the fermentation)
water	28 litres

The mixture is poured into sterilized plastic fermenting trays to a depth of approximately 3 cm, covered with two sheets of clean paper and stacked. The mixture is left undisturbed for 10–12 days and is fermented at 28–32 °C by bacteria (*Acetobacter aceti,* subspecies *xylinum*). When the right temperature is maintained the solid *nata de coco* will be 2.5 cm thick, but at lower temperatures the *nata* formation will be thinner.

The product is harvested and the cream on the lower surface scraped off with a clean knife. It is cut into approximately 1 cm cubes and soaked in several changes of water for at least one day to remove the sour vinegar taste. It is then boiled in water for five minutes and sugar is added (1 part sugar to 1 part *nata*), and the mixture is then left to stand overnight. The next day a small amount of water is added and the mixture brought to the boil. Flavourings and colourings may be added at this stage. The final product is packed into pre-sterilized jars with added syrup, capped and sterilized by immersing in boiling water for half an hour. It should have a clear crystalline appearance.

Poor quality *nata* results from undesirable micro-organisms in the fermenting mixture. Strict sanitary procedures must be followed since there is no heating of the *nata* substrate. The utensils and containers must be sterilized with boiling water before use, the processing area must be kept clean and workers must observe strict hygienic practices. Undesirable micro-organisms can also result in decreased growth, very soft *nata* or just a thin film of *nata* growth.

Bibliography

The reader is referred to the following texts on small-scale oil-seed and nut processing: Axtell and Fairman (1992), Head et al. (1995), Potts and Machell (1995) and UNIFEM (1993b).

Honey, syrups and sugar confectionery

Before the introduction of refined white sugar, honey, syrups and treacles made from plant saps were used as the main sweetening agents. Today they are commonly used to sweeten beverages and as ingredients for sugar confectionery. Confectionery has high sugar levels (78–84 per cent) which inhibit microbial activity, and products are stable for many months if processed and packaged correctly.

Honey and syrups

Honey

Honeybees deposit nectar into honeycombs and seal them with beeswax. Honey is a solution of sugars (glucose and fructose), vitamins, minerals and pollen, together with small amounts of volatile compounds that give its characteristic odour. It is twice as sweet as sugar (sucrose). The composition varies according to many factors, including the source of flower nectar, the time of year and weather at the time of pollen collection.

Syrups and treacle

Syrup and treacle are viscous liquids extracted from plant sources, including coconut and kitul palms, maple and sugar cane. Processing involves extracting the juice (sap) from the plant, and heating it to remove water and increase the sugar content. Table 10 outlines the various stages in processing and the equipment that is required for each stage.

Nutritional significance

These products are rich in sugar and therefore provide energy. They also contain vitamins and minerals, but the levels are too low to have any major nutritional significance. Honey is said to contain naturally occurring antibiotics which may have a positive effect on health.

Processing

Process notes

Extracting sap and honey

Sap is removed by 'tapping' the flower of kitul palms, the stem or fruit of coconut palms and the trunks of maple trees, and allowing the sap to drain into a collecting bowl. It is important that the sap is processed quickly after collection, to prevent fermentation. Collection bowls may be inadequately cleaned, causing a build-up of yeast, which will cause fermentation. Preservatives such as sodium benzoate and sodium metabisulphite can prevent fermentation, but with proper hygiene their use is unnecessary. Sugar cane is crushed with a roller mill to extract the juice.

Tapping

In honey extraction, the wax is first removed from the honeycombs using a special uncapping knife which may be heated by steam or simply dipped in hot water to melt the wax. There are several methods by which the honey can be extracted from the honeycombs: they may be crushed by hand, but this makes removal of comb fragments difficult, and the resulting honey has a lower quality. Honey may be drained from combs through stainless steel or plastic sieves into a stainless steel container over approximately three days. Although this method is effective, it may allow the product to be contaminated by insects. The combs can be placed in a frame in a manual or a powered honey centrifuge (Directory sections 1.1 and 6.1). The spinning action throws honey from the

Table 10. Various stages in the processing of honey and syrups

Processing stage	*Honey*	*Syrups*	*Treacle/ molasses*	*Equipment*	*Directory reference*
Extraction	✓	✓	✓	Honey knife Honey centrifuge Sugar cane rollers	1.1 1.1, 6.1 23.2
Filtration	✓	✓	✓	Filter cloth Stainless steel mesh Strainers	18.0 18.2 18.3
Conditioning	✓			Fan	–
Boiling		✓	✓	Boiling pan Steam jacketed pan Heat source Thermometer	27.0 27.1 25.0 38.5
Packaging	✓	✓	✓	Filling machines Capping equipment	17.1 26.2

combs and it is collected inside the centrifuge.

Filtration
Honey, syrups, and treacle contain pollen seeds, bee hairs, and vegetable matter that is filtered out using a muslin cloth or a fine stainless steel mesh. They are easier to filter if they are warmed to 26–30 °C so that they flow more easily.

Honey conditioning
Conditioning removes excess moisture and concentrates the sugar to 82 per cent by either gently heating the honey in an open pan at a temperature of no more than 35 °C or blowing air at 25–40 °C over the surface of the honey with a fan. There are many drawbacks to heating: there is little temperature control, the product requires constant stirring to avoid localized overheating, darkening and production of HMF[1] or off flavours in the product if the temperature gets too high. The sugar content can be checked using a refractometer. Most refractometers calculate sugar concentration on the basis of sucrose but honey consists of fructose and glucose, and a specifically designed honey refractometer is needed (Directory section 38.4).

Tapping a kitul palm

Heat treatment
Syrups and treacle are heated to reduce the water content, and the temperature and time

1 Hydroxy methyl furfuraldehyde is an important quality indicator and is produced by a reaction of sugars and acids in honey above 20 °C. Despite its somewhat alarming name it is not dangerous to health. High levels of HMF indicate that the honey is either old or has been heated or stored at a high temperature.

of heating should be carefully controlled. If the sugar content is too low (i.e. below 75 per cent), fermentation can occur. Heating also caramelizes the sugars, giving the product a darker colour. It is preferable to use a steam jacketed pan, which although more expensive, is better able to avoid localized overheating and maintain the quality of the product. A refractometer (Directory section 38.4) is used to measure the sugar concentration.

Packaging

Honey and syrups can be sold directly in bulk containers (especially for export) such as 300 kg drums that are internally lacquered with a food-grade material or coated with beeswax. Honey that is fresh from the hive is stable, but will quickly absorb moisture from the atmosphere, and drums should be completely filled to exclude moisture and air, which can react with the honey and cause oxidation. They can also be packaged in securely sealed glass jars/bottles to prevent leakage and contamination by insects.

Honey on a honeycomb

Adding value to honey

Processing by-products may add considerable value to honey. The formulations below are taken from an article produced in *Beekeeping and Development* and were originally developed by Elaine White. Her book *Super Formulas* (White, 1993) should be essential reading for those interested in this subject.

Wax for coating cheese

Many cheeses are externally waxed to prevent them drying and to retard mould growth.

Take 337.5 g beeswax and 62.5 g vegetable shortening, and heat ingredients together in an oven at 120 °C, checking the temperature with a thermometer. Hold down the cheese under the hot wax for 10 minutes; this kills most surface micro-organisms. Remove the cheese and allow the wax to cool to 70 °C. Briefly dip the cheese in the wax and repeat until it is covered with a wax layer about 1.5 mm in thickness.

Petroleum jelly

Slowly melt 25 g of beeswax in a double boiling pan. When it is liquid, stir in half a cup of baby oil. Pour into attractive containers.

A balm for dry lips

Take:

shredded beeswax	1 tbsp
petroleum jelly	1 tbsp
honey	1 tsp
lanolin	1 tbsp

A few drops of aromatic essential oil (peppermint, eucalyptus, wintergreen or camphor).

Melt the wax, lanolin and petroleum jelly in a double boiler. Add the honey and essential oil. Stir the mixture until cool.

Sugar confectionery

Sugar confectionery includes boiled sweets, toffees, chocolates, fudges, marshmallows and fondants. These are non-essential for health, but are consumed by people from most income groups, sometimes at special occasions or ceremonies and sometimes as a treat or reward.

Nutritional significance

The main ingredient is sugar (sucrose) which contributes energy to the diet. There is a danger that if sweets are consumed in excess over a prolonged period they may contribute to obesity or tooth decay.

Processing

It is possible to make a wide variety of products from sugar by varying the ingredients, temperature of boiling, and the method of shaping the product. The sequence of production is the same for each product: mix together the ingredients, boil the mixture to the required temperature, cool, shape and pack. The factors that affect the quality of sweets are:

- the degree of sucrose inversion
- the time and temperature of boiling
- the residual moisture content
- addition of other ingredients.

Degree of sucrose inversion

Sweets that contain high concentrations of sugar may undergo crystallization (or 'graining') during manufacture or storage. Although this may be desirable for some products (such as fondant and fudge), in most other cases it is seen as a quality defect. When a sugar solution is heated, a percentage of sucrose breaks down to form 'invert sugar' which inhibits sucrose crystallization and increases the overall concentration of sugars in the mixture. However, it is difficult to accurately assess the degree of invert sugar produced by simple heating. Cream of tartar or citric acid breaks down sucrose into invert sugar and may be used to control the amount of inversion. A more accurate method of ensuring the correct balance of invert sugar is to add glucose syrup, as this will directly increase the proportion of invert sugar in the mixture.

The amount of invert sugar must be controlled – too much may make the sweet absorb water from the air and become sticky; too little will be insufficient to prevent crystallization. Ten to fifteen per cent invert sugar is sufficient to give a non-crystalline product.

Time and temperature of boiling

The time and temperature of boiling are very important as they directly affect the final sugar concentration and hence the moisture content of the sweet.

At a given concentration of sugar, boiling occurs at the same temperature (assuming the same altitude above sea level), and therefore each type of sweet has a different heating temperature (see Tables 11 and 12).

Table 11. Boiling point of sucrose solutions

Sucrose concentration (%)	*Boiling point (°C)**
40	101.4
50	102
60	103
70	105.5
75	108
80	111
85	116
90	122
95	130

Note: *at sea level

Table 12. Boiling points for different types of sweet

Type of sweet	*Temperature range of boiling (°C)*
Fudge	116
Fondants	116–121
Caramels and regular toffee	118–132
Hard toffee (e.g. butterscotch)	146–154
Hard-boiled sweets	149–166

An accurate way of measuring the temperature is to use a sugar thermometer (Directory section 38.5). Other tests involve removing a sample of boiling mixture, cooling it in water and examining it when cold. The textures of the products are known by distinctive names such as 'soft ball', 'hard ball' etc.

Moisture content

The moisture in a sweet influences its packaging requirements and storage behaviour, and determines whether the product will dry out or pick up moisture. Sweets that contain more than 4 per cent moisture are likely to crystallize at the surface on storage, later spreading throughout the sweet. Added ingredients also affect the shelf life. Toffees, caramels and fudges, which contain milk solids and fat, have a higher viscosity which controls crystallization, but may also make the sweet prone to rancidity and shortened shelf life.

Table 13 outlines the processing stages for a selected range of confectionery items.

Process notes and quality assurance

Boiling

There are three types of equipment used for boiling: an open boiling pan, a steam jacketed pan or a vacuum cooker. Steam jacketed pans (Directory section 27.1) are fitted with scrapers and blades which make mixing and heating more uniform, and lessen the possibility of localized overheating and burning onto the pan wall. Vacuum cookers are not used at a small scale as they are very expensive.

Cooling

All sweets are cooled slightly before being shaped by pouring the boiled mass onto a metal, stone or marble table to cool the product uniformly. The table should be clean and free from cracks, as they may harbour dirt and micro-organisms. It is important that the boiled mass is cooled sufficiently so that there is no danger that the operator may suffer burns.

Beating

Beating controls crystallization and produces crystals of a small uniform size. For example, in the production of fudge the sugar mass is poured onto a table, left to cool and then beaten with a wood or metal beater.

Forming/setting

There are two main ways of forming sweets: cutting into pieces, or setting in moulds. Moulds may be a greased and lined tray, made from rubber, plastic, metal or wood, or indentations in starch. It is possible to make starch moulds by preparing a tray of cornflour (cornstarch), not packed too tightly, and making impressions in the starch using metal or wooden shapes. The mixture is poured into the impressions and allowed to set.

Pulling toffee to incorporate air

Packaging

When sweets are stored without packaging, the high sugar content absorbs moisture from the air making the sweet sticky and grainy, especially in areas of high humidity. Packaging in individual wrappers of waxed paper, aluminium foil, polypropylene or cellulose film can be done by hand, but for higher production levels, semi-automatic

Packaged sweets

Table 13. Processing stages for different types of confectionery

Processing stage	*Hard-boiled sweets*	*Fondants*	*Toffees/ caramel*	*Fudge*	*Jellies*	*Marsh-mallows*	*Equipment*	*Directory reference*
Mix ingredients	✓	✓	✓	✓	✓	✓	Weighing and measuring equipment	39.0
Boil	✓	✓	✓	✓	✓	✓	Heat source Boiling pan Steam jacketed pans Thermometer	25.0 27.0 27.1 38.5
Cool	✓	✓	✓	✓	✓	✓		
Beat		✓		✓	✓	✓	Hand whisk or liquid mixer	24.0
Form/set	✓	✓	✓	✓	✓	✓	Moulds Starch moulding Cutting equipment	4.2 10.0
Pack	✓	✓	✓	✓	✓	✓	Heat-sealer Wrapping equipment	26.1 26.3

wrapping machines are available (Directory section 26.3). For further protection the individually wrapped sweets may be packed in a heat-sealed polythene bag, glass jars or tins with close-fitting lids.

Types of sweets

Fondants and creams

Fondant is made by boiling a sugar solution with the optional addition of glucose syrup. The mixture is cooled and then beaten to control the crystallization process and reduce the size of crystals. Creams are fondants which have been diluted with a weak sugar solution. These products are not very stable because of their high water content, and have a shorter shelf life than many other confectionery products. Both fondants and creams are commonly used as soft centres for chocolates and other sweets.

Gelatin sweets

These sweets include gums, jellies, pastilles and marshmallows. They are distinct from other sweets as they have a spongy texture which is set by gelatin. An example of a jelly (or jujube) is as follows:

Ingredients	
Sugar	250 g
Gelatin	25 g
Liquid glucose	75 g
Citric acid	2 g
Essence	5 drops
Colouring	5 drops
Water:	100 ml (½ cup) to dissolve gelatin
	100 ml (½ cup) to boil sugar

Method

Sugar and liquid glucose are dissolved in water and boiled to 120 °C. The mixture is cooled slightly and a solution of gelatin dissolved in warm water is added. Citric acid, flavours and colours are added, and the jelly is poured into oiled moulds and allowed to set. The sweets are then cut into pieces, coated with sugar and wrapped.

Toffee and caramels

These are made from sugar solutions with the addition of milk solids and fats. Toffees have a lower moisture content than caramels and consequently have a harder texture. As the product does not need to be clear, it is possible to use unrefined sugar such as jaggery or gur, instead of white granular sugar. An example of a toffee recipe is as follows:

Ingredients	
Sugar	500 g
Coconut	1
or	
Coconut milk	100 ml (1 cup)
or	
Milk powder	100 g dissolved in 200 ml water
Peanut/cashew	25 g
Cardamom	5 seeds
Vanilla	as required
Colouring	as required

Method

Sugar and milk (coconut or milk powder) are dissolved in water and boiled to 130 °C. Other ingredients are stirred into the mixture, and boiling is continued for a short while to a temperature of 132 °C. The toffee is spread on a tray lined with oiled paper, cooled and cut into pieces before packing.

Peanut bars

Use the following ingredients:

peanuts	2.5 kg (shelled, roasted and ground)
sugar (brown)	5 kg
water (optional)	2.5 kg
vitamin E (optional)	10 g

Process flow sheet for peanut brittle

Stage	*Equipment required*	*Directory section*
Peanuts (groundnuts) ↓		
Shell ↓	Peanut sheller	12.0
Roast ↓	Oven Peanut roaster	25.0 32.1
Grind to coarse pieces ↓	Pestle and mortar or coarse mill	23.0
Mix with syrup ↓		
Boil ↓	Boiling pan Double jacketed pan Heat source Thermometer	27.0 27.1 25.0 38.5
Cool ↓		
Form ↓		
Cut ↓	Cutter	10.0
Pack	Heat-sealer	26.1

Process notes

Raw materials

Peanuts are prepared by shelling (manually or with a peanut sheller) and roasting until slightly brown and aromatic. Alternatively they can be prepared by boiling in water into which vitamin E has been added to delay the onset of rancidity or off-flavour development. The cooked kernels are then oven-dried.

Grinding

Peanuts are ground with a pestle and mortar or a coarse mill into smaller, coarse pieces.

Syrup

A syrup is prepared by heating sugar with water. It is heated without stirring until thick.

Mixing

The ground peanuts are added to the syrup and stirred constantly to avoid scorching.

Boiling
The mixture is either heated to 150 °C or tested to see if it is ready by dropping a sample into cold water, when it should form into hard balls.

Forming
It is then poured onto a greased board, rolled to a sheet 0.5 cm thick, cooled and cut into bars of convenient length.

Packaging
The peanut bars are wrapped with cellophane or wax-coated paper.

Quality assurance
The main points are as follows.

Raw materials
Only good quality peanuts, free from mould and rancidity, should be used. Broken kernels, as long as they are not infected or rancid, can be used for this product.

Roasting
The quality of nuts in the final product is determined by the time and temperature of roasting and the size and moisture content of the nuts.

Syrup
The colour and texture of the final product is determined by the heating of the sugar and the ratio of the nuts to sugar. The hardness/crispiness of the product depends on the amount of heating and the amount of nut pieces added to the mixture. Care has to be taken to produce a hard crispy product, while ensuring that the syrup mixture does not burn.

Packaging
The peanut brittle will stay hard and crunchy only if it is packaged in airtight, moisture-proof wrappers and stored in a cool dry place.

Hard-boiled sweets
These are boiled and then cooled to form a solid mass containing less than 2 per cent moisture. There is a wide range of potential products by the use of different flavourings and colourings and moulding into different shapes.

Further reading
See Agrodok (1991); Commonwealth Secretariat International Bee Research Association (1979); Krell (1996); White (1993).

Meat and fish products

Meat

In many developing countries meat is eaten fresh and, with the exception of dried meats, processed products are not commonly available or in significant demand. A wide range of animals are used as a source of meat, from domesticated cows, pigs and chickens to camels, horses, hamsters and wild animals such as deer, snails, boars and monkeys. The amount and type of meat eaten in a particular country are determined by factors such as cost, availability and cultural or religious acceptability.

The conditions under which animals are slaughtered are often unhygienic, and meat may be displayed for sale in the open air without any covering. This enables bacteria to be transmitted by flies, other animals and birds. Meat, like fish and milk, is a low-acid, moist food which provides a good environment for the growth of these bacteria. This may lead to rapid deterioration in tropical climates, and fresh meat therefore has a very short shelf life. Harmful micro-organisms may also grow on meat and produce food poisoning when eaten. This, together with infectious organisms such as parasites which grow in the meat, makes proper handling and preparation of meat products essential. In this chapter the focus is on processed meat products and not the sale of fresh meat for direct consumption. However, fresh meat is the raw material for processing, and a note on hygiene and handling is therefore included below.

Unlike other commodity groups in this book, there are a limited number of processed products made from meat, and for this reason only a few examples (sausages, dried meat and burger patties) are included. It is strongly recommended that meat products should be made only by experienced food processors, because of the potential risks from food poisoning.

The two main methods of processing meat products are

1. salting, smoking and/or drying, and
2. grinding to form minced meat, sausage-meat or pâté.

In this chapter, one example from each group is described (*biltong* as an example of a dried, salted meat, and sausages as an example of a ground meat product). It should be noted also that there are hundreds of types of sausages, many of which are also smoked and/or dried to aid preservation. Further details are given in specialist publications (Hippisley Coxe and Hippisley Coxe, 1994).

Nutritional significance

Meat is a good source of easily digestible protein and contains essential amino acids which are vital for growth and maintenance of the body. It is also a good source of vitamins and minerals, particularly iron. Processing does not have a substantial effect on the nutritive value when compared to normal cooking processes.

Fish

Small-scale fisheries in developing countries provide nutritious food that is often cheaper than meat and therefore available to a larger number of people. In many countries the bulk of fish is sold fresh and processing is either to supply distant markets or to produce a range of products with different flavours and textures. Fish is an even more perishable food than meat and most fish becomes inedible within 12 hours at tropical temperatures. Spoilage begins as soon as the fish dies, and processing should therefore be carried out quickly to prevent the growth of spoilage bacteria. Fish is also a low acid food and is therefore very susceptible to the growth of food-poisoning bacteria. Lowering the temperature using ice or refrigeration preserves the fish, but causes no noticeable changes to the texture and flavour. Other methods of preservation cause changes to the flavour and texture of the fish, producing a range of different products. These include:

- cooking (for example, boiling or frying)
- lowering the moisture content (by salting, smoking and drying)
- increasing the acidity (by fermentation).

Nutritional significance

Fish is a good source of high quality protein and contains many vitamins and minerals. White fish, such as haddock and seer, contain very little fat (usually less than 1 per cent), whereas oily fish, such as sardines, contain between 10 and 25 per cent fat and a range of fat-soluble vitamins (A, D, E and K) and essential fatty acids.

Meat processing

Fresh meat should be kept under refrigeration or cool storage and covered to protect it from insects and animals. Additionally, the hands and clothes of workers who handle meat should be regularly cleaned. In all meat processing, the aim is twofold: to preserve the meat for a longer storage life, and to change the flavour and texture to increase variety in the diet.

Salted, smoked and dried meat

In drying, salting and smoking, the main aim is preservation, which is achieved by reducing the moisture content of the meat. In salting, high concentrations of salt also inhibit most micro-organisms. Processing involves either rubbing salt into the meat (salting) or soaking in salt solution (curing or brining). In smoking, the meat is dried and sometimes partially cooked by heat from the fire. Chemicals in the smoke preserve the meat and smoke also adds distinctive and attractive flavours and colours to the meat.

Biltong production

Dried, salted and smoked meats are found widely in Africa where they are important traditional foods. Possibly the best known example is *biltong* which is used as a snack food in Southern Africa. It is made from strips of dried, salted meat which are dark brown with a salty taste and a flexible, rubbery texture. It is mainly used as a snack and has a shelf-life of several months under correct storage conditions.

Production of biltong

Biltong is produced by removing the fat and cutting the meat into thin strips. These are either soaked overnight in a mixture of salt and herbs/spices, or these ingredients are rubbed into the meat. It is then sun-dried by

Process flow sheet for biltong

Stage	Equipment required	Directory section
Wash meat		
↓		
Trim the fat from the meat	Knife	10.2
↓		
Slice	Cutting and slicing equipment	10.2
↓		
Rub salt and flavourings into the slices	Weighing equipment	39.0
↓		
Dry	Mats, racks	
	Solar dryer	14.1
	Fuel-fired dryer	14.2
↓		
Pack	Traditional packaging materials	26.7
	Heat-sealing machine	26.1

hanging the strips on a frame under an insect-proof mesh. If *biltong* is to be produced in more humid or cooler climates, a dryer could be used instead of sun-drying.

The principle of preservation is to inhibit enzyme and microbial action by addition of salt to the surface of the meat and removal of moisture by drying. When used, potassium nitrate/sorbate also has a preservative effect.

Process notes

Fresh meat
Good quality beef from the hindquarters should be used. It must be kept cool, handled in a hygienic way and protected from flies, dust and dirt.

Separate fat
Remove as much fat as possible with a knife.

Slice
Use a sharp sterilized knife to cut the meat into strands along the muscle fibres and then across the muscle fibres to produce uniform sized strips 2 cm wide, 1 cm thick and 20–25 cm long. For more precise slicing, electrically powered knives are available. Discard all fat and connective tissue.

Salt/spices
Usually a mixture of salt and spices is prepared and rubbed onto the pieces of meat. Typical spice formulation (for 100 kg meat):

salt	3.74 kg
sugar	1.87 kg
potassium nitrate	0.02 kg
potassium sorbate	0.2 kg
mixed spice	0.21 kg
black pepper	0.10 kg
onion powder	0.03 kg
garlic powder	0.03 kg
ground ginger	0.03 kg
mustard powder	0.03 kg

Alternatively the meat is soaked in a solution of these ingredients.

Dry
Hook each slice of meat onto a hook and hang the hooks on hanging wire. Sun-dry the meat for 7 to 10 days depending on the climate (e.g. 25–30 °C, <80 per cent humidity, with a gentle breeze). Enclose the meat in netting or gauze to protect it from flies while it dries.

Pack
Salted, dried meat needs to be protected from moisture pick-up and attack by insects. If the climate is dry it may be packaged in traditional materials such as jute bags and cane baskets. For more humid areas, it is packed in polythene bags (e.g. 50 g net weight) and heat-sealed.

Store
In cool, dry conditions away from sunlight.

Quality assurance

The main points are as follows.

Hygiene

It is essential that all tools, equipment and surfaces are thoroughly cleaned and sterilized in boiling water for 10 minutes before and after processing. Workers' hands must be washed thoroughly with detergent or soap and clean water. As *biltong* is not cooked before eating it is a potentially serious health hazard. Strict personal hygiene and hygienic food handling practices should be enforced to prevent food-poisoning bacteria from contaminating the product.

Raw materials
Meat should be freshly slaughtered, free from disease and handled carefully to prevent contamination. Meat that can be easily cut into strips is best, but cheaper parts of the animal can also be used. Shin meat does not produce good *biltong*. The meat should be boned and trimmed of excess fat and any tendons. Spices should be from a reputable supplier as these are often a source of food-poisoning bacteria.

Drying
The meat should be protected from insects and animals during drying. The main quality factors are colour, taste and texture of the product. Colour and texture are determined by the drying conditions and taste is mostly determined by the amount of spices used.

Processing
The main control points are as follows.

- size of the strips, which should be uniform to give similar drying times
- drying rate, which affects the product quality and moisture content; this depends mostly on the temperature and air speed of the drying air and size of the strips
- amount of salt and spices rubbed into the meat, which prevents surface bacterial growth during the initial stages of drying and contributes to the distinctive flavour of the product.

Quanta is a similar dried and spiced beef product used as snack food in North Africa. Spices are mixed with fermented honey, and about 120 g spice (2.5 per cent salt, 1.5 per cent black pepper, 10 per cent spiced chilli pepper powder, with 86 per cent flour) is added to each kilogram of meat and mixed in manually. The evenly spiced strips are hung on a string suspended in a well-ventilated, dust-free area and left to hang for five to seven days until they break with light hand pressure or are crunchy upon chewing.

Salted meat

In Asia chunks of pork or other meats such as beef or chicken are salted and later cooked after thoroughly washing out the salt. Pieces of meat are dipped in a saturated salt solution and then pressed, rubbed with salt crystals, wrapped tightly in cloth, tied closely with a rope and hung. Dipping in salt solution helps prevent microbial spoilage, and the salt also draws out water from the meat. Subsequent pressing and wrapping removes further water and hanging enables any remaining water to drip away. The product is therefore dry and has a high salt content, which acts as a preser vative. It can be kept for about a month, but rancidity and spoilage gradually develop. In North Africa a similar salted pork product is made by dry salting meat scraps or pork feet, sometimes with the addition of spices. It is common to use very fatty meat cuts that are difficult to sell, and red food colouring is also sometimes used to colour the product. The process involves heavy dry salting of the meat: liberal amounts of fine salt granules are rubbed onto the meat and extra salt is sprinkled to cover the meat, which is then packed in covered drums. Initially the meat may need to be weighed down with heavy objects such as cleaned stones prior to formation of the pickle. The drums are kept covered for at least seven days, during which time meat juices are released and form a strong brine to submerge the meat, protecting it from air. Salt levels may range from 12 to 18 per cent in the final product, and it is therefore stable up to three or four months. When it is required for a meal the product is boiled for about one hour to soften the tissue and de-salt it before incorporating it into the soups.

Smoked meat

In West Africa this product is made from pork shoulders, feet, head or offal, which are hot-smoked at temperatures of about 100–120 °C and added to soups or stews. Other meats, including venison and grasscutter, are also used. The products have a well-developed smoky flavour but are only partially cooked. Only mild salting is carried out (e.g. salt at 3–5 per cent of the weight of meat is rubbed into the surface). It is stable for about four days when packaged in polythene. If it is refrigerated it has a shelf life of about two weeks at 5–10 °C. Smoking is done on a grill mounted on a tripod arrangement. The meat is turned regularly for 6 to 10 hours to ensure uniform heating and smoking. Good quality fuelwood is used and sugar-cane bagasse may also be used to generate smoke. This material imparts a finer smoked flavour and an attractive golden finish to the meat. Hot-smoking must be carefully done to avoid overcooking and charring the product. In other areas slightly salted pork loins, bellies, etc. may be cold-smoked for specific markets. The cold-smoking involves temperatures of around 40 °C to produce partial cooking but not to the extent of hot-smoking.

Meat smoking over an open fire

Ground meat products

The process of grinding meat into small pieces such as minced meat or pastes (sausage meat) aims to change the texture of the meat and allow it to be formed into different shapes. Most commonly, these are cylindrical sausages enclosed by a casing or skin, or flat discs (patties). Both enable more rapid cooking, and both methods allow for spices and other flavours to be included. The grinding process does not help to preserve meat and in fact causes more rapid spoilage. This is because there is more chance for bacteria from workers' hands or dirty equipment to become mixed with the meat. Grinding also releases enzymes from the

Process flow sheet for sausages and patties

Stage	*Equipment required*	*Directory section*
Wash meat		
↓		
Trim the fat from the meat	Knife	10.2
↓		
Mince/chop	Cutting and slicing equipment	10.2
	Bowl chopper	10.2
	Mincers and shredders	10.3
↓		
Add ingredients	Weighing equipment	39.0
↓		
Form/mould or extrude into casing (skin)	Filler (sausage stuffer)	17.2, 37.0
↓		
Pack	Heat-sealing machine	26.1

meat which causes changes to its flavour and texture. It is therefore essential that operators must be thoroughly trained in hygienic processing, all equipment should be thoroughly sterilized and all processing should be done quickly, preferably at a low temperature (below 10 °C) using ice or refrigeration to slow down bacterial growth. Preservation is achieved by adding chemical preservatives, refrigeration or freezing, smoking, drying or pasteurization.

Suitability for small-scale production

The potential hazards of infection, food poisoning and product deterioration mean that only people with knowledge and experience of these hazards during routine production should consider becoming meat processors.

Process notes

Fresh meat

Good quality meat should be used. It must be kept cool, handled in a hygienic way and protected from flies, dust and dirt.

Mincing

The cheapest option is to use a knife, but for more thorough and efficient chopping, bowl choppers are recommended. They are specifically designed for chopping meat into a homogenous sausage emulsion and consist of a rotating, horizontal bowl and a set of rotating vertical knives. Manual or powered mincers/grinders may be used to produce minced meat for burger patties. They consist of a screw shaft inside a metal barrel, and a cutter/die assembly at the outlet. As the handle is turned, the screw rotates and the minced meat is pushed along the barrel chopped by the knives and forced out through the die.

Forming/moulding

For burger patties the mixture is weighed into portions and moulded and pressed by hand. However, although this method is effective, it gives an opportunity for bacteria from workers to contaminate the patty. Burger moulds are available or can be easily fabricated from local materials.

Filling
Sausage skins are filled using a powered or hand-operated sausage stuffer (having a piston type plunger, which may be fitted with a range of different sized nozzles).

Packaging/storage
Simple paper or polythene wraps are sufficient to keep the products clean and prevent contamination by insects and dirt. They should be kept under refrigeration (<10°C) to slow down the actions of enzymes and micro-organisms and give a shelf life of one to seven days. These products may also be frozen to give a shelf life of up to six months.

Quality assurance
The main points are as follows.

Hygiene
It is essential that all tools, equipment and surfaces are thoroughly cleaned and sterilized in boiling water for 10 minutes before and after processing. Strict personal hygiene and hygienic food handling practices should be enforced to prevent food-poisoning bacteria from contaminating the product. The main quality factor is the fineness of the texture of the product.

Raw materials
Meat should be freshly slaughtered, free from disease and handled carefully to prevent contamination.

Chopping
It is important to control the size of the particles of meat, since this determines the texture of the product. This is controlled by the method of shredding/chopping.

Storage
Meat products should be stored in a refrigerator, at temperatures below 10 °C. If facilities are available, the sausages/patties can be frozen to extend the shelf life up to six months.

Fish products

Cooked fish
Cooking allows short-term preservation of fish for a few days before deterioration becomes noticeable. A range of methods is used to cook fish but all have the same principle: enzymes are inactivated and many of the bacteria present on the surface of the fish are killed by heat. Boiling and poaching involve cooking fish in hot water, and frying uses hot oil. These techniques are simple and require only basic household equipment, which makes them suitable for small-scale production. Cooked fish products are mostly consumed immediately and therefore do not require sophisticated packaging. They should be kept hot (above 60 °C) until consumed. The shelf life can be extended for a few days using refrigerated storage, but extreme care should be taken to prevent recontamination of cooked products by food-poisoning bacteria.

Cooled/frozen fish
Any reduction in the fish temperature prior to processing will maintain the quality for longer. Fish can be kept cool by covering it with clean, damp sacking and placing it in the shade. This method is simple and requires no special equipment, but the fish still begins to deteriorate within a few hours. A more effective method is to pack the fish in boxes, covered with ice, although the cost of the ice may be greater than the value of the fish in some countries. Freezing is another alternative for long-term preservation, but it is expensive in terms of equipment and operating costs.

Cured fish products
Curing involves drying, dry salting or brining (soaking in salt solution) or smoking. These may be used alone or in various combinations to produce a range of products with a long shelf life. Table 14 indicates the processes for a range of fish products.

The objective of each processing method is

Table 14. Stages in fish processing

Process/product	*Gut*	*Wash*	*Salt/brine*	*Dry*	*Smoke*	*Pack*
Dried fish	✓	✓		✓		✓
Smoked fish	✓	✓			✓	✓
Dried and salted fish	✓	✓	✓	✓		✓
Dry-salted and smoked fish	✓	✓	✓	✓	✓	✓
Brined and smoked fish	✓	✓	✓		✓	✓

to reduce the moisture content in the fish and prevent the growth of spoilage micro-organisms.

Dried fish for sale in the market

Dried fish

In drying, heat and movement of air remove moisture. To prevent spoilage, the moisture content should be reduced to 25 per cent or less, depending on the oiliness of the fish and whether it has been salted. Traditionally, whole small fish or split larger fish are spread in the sun on mats, nets, roofs o raised racks. However, there is little contro over drying times and the fish is exposed t insects or vermin and contamination by san or dust. Solar or artificial dryers (Director sections 14.1, 14.2) are alternatives to sun drying, and there has been a great deal o research and development to increase dryin temperatures and drying rates to produce lower moisture content in the final product and improvements to fish quality. Combination dryers that combine sun and artificia heat are useful for drying fish. Fish ar placed on trays, which are placed in the su to dry. At night, or when it is cloudy or raining, the trays are stacked over a simple heating compartment, a roof and chimney ar placed on top and drying continues by direc heating. Both solar and artificial dryer attempt to improve control over drying conditions and produce a better quality product However, it should be pointed out that it i only advantageous to use such dryers if ther is a market for a higher-quality product or i the fish would otherwise be lost.

Salted fish

A concentration of 6–10 per cent salt in fis tissues will prevent growth of food-poisoning bacteria, and a product preserved by salting will have a longer shelf life. However, group of micro-organisms known a 'halophilic' (salt-loving) bacteria will spoi salted fish even at the recommended sal concentration. Further removal of the wate by drying is needed to inhibit these bacteria

During salting or brining two processe

take place simultaneously: water moves from the fish into the solution outside and salt moves from the solution into the flesh of the fish. Salt can be applied by rubbing it into the fish flesh or by making alternate layers of fish and salt (recommended levels of salt are 30–40 per cent of the weight of fish). The salt must be uniformly applied to ensure that the concentration in the flesh is sufficient to preserve the fish. Brining involves immersing the fish into a solution of 16–18 per cent salt. The advantages of this method are that the salt concentration can be more easily controlled and salt penetration is more uniform. Brining is often used in conjunction with drying.

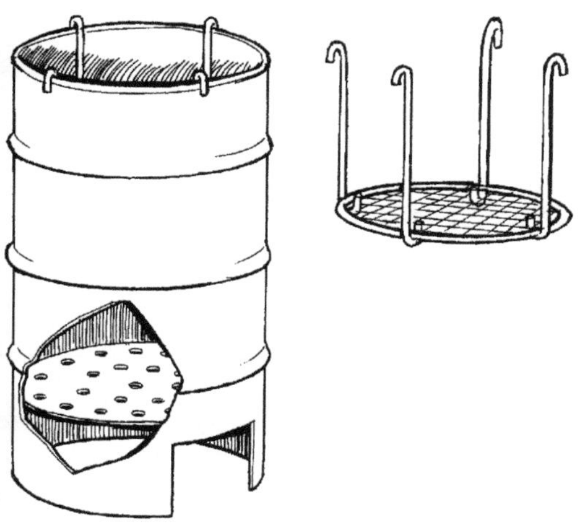
Oil drum smoker

Smoked fish

The preservative effect of smoking is due to drying and the deposition onto the surface of the fish of natural chemicals in wood smoke which inhibit bacteria. Heat from the fire dries the fish, and if the temperature is high enough the flesh also becomes cooked. As a general principle, the longer it is smoked the longer its shelf life will be. Smoking can be either *cold-smoking* – where the temperature is not high enough to cook the fish (less than 35 °C) or *hot-smoking* – where the temperature is high enough (above 60 °C) to cook the fish.

Hot-smoking is often the preferred method because the process requires less control than cold-smoking and the product shelf life is longer. Hot-smoking has the disadvantage of greater fuel consumption. There are various types of smoking kiln used in different parts of the world. Although traditional kilns have low capital costs, they commonly have an ineffective air-flow system, which results in poor economy of fuelwood and lack of control over the temperature and smoke density. Improved smokers include the oil drum smoker and the Chorkor smoker (Directory section 33.0).

Chorkor smoker

Process notes

Raw material

Use fresh fish, and wash in clean water. Remove the guts and wash again to remove all traces of blood. Small fish are usually left whole, while larger fish are cut open or cut into steaks for smoking.

Salt

Salting is optional and is dependent upon local taste and the availability of salt. Salt can be rubbed into the skin of the fish before smoking.

Smoking

Fish are laid flat on their sides on platforms above a fire, or to increase the holding

Process flow sheet for smoked fish

Stage	*Equipment required*	*Directory section*
Wash fish ↓		
Gut ↓	Knife	10.0
Wash ↓		
Salt (optional) ↓	Weighing equipment Brine meter	39.0 38.7
Smoke ↓	Smoking equipment	33.0
Dry ↓	Mats, racks Solar dryer Fuel-fired dryer Combined dryer	 14.1 14.2 14.2
Pack	Heat-sealing machine	26.1

capacity and the flow of smoke they may be packed vertically with their heads down over the fire. The longer the fish is smoked, the drier it becomes and the more suitable it is for long-term storage.

The main control points are as follows.

- size of the fish, which should be uniform to give similar smoking and drying times
- time and temperature of smoking and type of fuel used
- drying rate, which affects the product quality and moisture content; this depends mostly on the temperature and air speed of the drying air and size of the fish.

Packaging

Packaging of cured fish products should prevent moisture pick-up and recontamination by insects and micro-organisms. Traditional packaging materials include cane baskets, leaves and jute bags. Alternatives include polythene bags or wooden and cardboard packs. Two may be combined, as in a polythene bag enclosed in an outer cardboard pack.

Quality assurance

The main points are as follows.

Hygiene

Strict personal hygiene and hygienic food handling practices should be enforced to prevent food-poisoning bacteria from contaminating the product. The fish should be protected from insects and animals during smoking and drying.

Smoking

The main quality factors are colour, taste and texture of the product. Colour, taste and texture are determined by the smoking conditions and the amount of salt used.

Salt
The amount of salt used to prevent surface bacterial growth during the initial stages of processing, which also contributes to the distinctive flavour of the product.

Fermented fish

Fermentation is a process by which beneficial lactic acid bacteria increase the acidity of the fish, and prevent the growth of spoilage and food-poisoning micro-organisms to extend its shelf life. The bacteria also soften the flesh and alter the flavour of the fish. There are many types of fermented products depending on the extent of fermentation. For example:

- fish which retains its original texture
- pastes
- liquids/sauces.

There is little specialized equipment required and the process may easily be carried out on a small scale. The flow sheet below outlines stages in the production of a fermented fish paste.

Production stages for fish paste (bagoong)

This East Asian product is made from whole or ground fish, fish roe or shellfish. It is reddish-brown in colour, although this will depend on the raw materials used, and is slightly salty with a cheese-like odour.

Process notes

Raw materials

Good quality, fresh fish should be used. Small fish are often left whole; larger fish may be cut into smaller pieces. The guts are removed and the fish washed in clean water.

Salting
Fish are placed in a wooden vat and salt is added at a ratio of 3 kg fish to 1 kg salt.

Fermentation
The salt and fish mixture is transferred to earthenware jars or oil drums and covered with cheesecloth for five days. After this time the drums are sealed and left in the sun for seven days. After this time the mixture is transferred into 20 litre cans, which are left to stand for 3 to 12 months to allow for further fermentation of the product.

Process flow sheet for fish paste (bagoong)

Stage	*Equipment required*	*Directory section*
Wash fish		
↓		
Gut	Knife	
↓		
Add salt (approx. 5%)	Weighing equipment	39.0
↓		
Ferment (for several months)	Fermentation bin	
↓		
Add colouring (optional)		
↓		
Pack	Heat-sealing machine	26.1

Packaging
Glass bottles are often used for the better-quality products, but earthenware pots and even plastic bags are also used.

Quality assurance

The main points are as follows.

Hygiene

It is essential that all tools, equipment and surfaces are thoroughly cleaned and sterilized in boiling water before and after processing. Strict personal hygiene and hygienic food handling practices should be enforced to prevent food-poisoning bacteria from contaminating the product.

Raw materials

Fish should be fresh, free from disease and handled carefully to prevent contamination.

Fermentation
Quality of the product depends on the following main factors:

- amount of salt added to the fish
- the temperature and time of the fermentation
- using an airtight fermentation bin.

Fermentation bins
Earthenware pots, metal or food-grade plastic oil cans, drums and glass containers are each used, provided that they are airtight.

Further reading

See Caribbean Development Bank (1987); Clucas and Ward (1996); ILO (1985a, 1985b); Johnson and Clucas (1996); Lee et al. (1993); Norman and Corte (1985); UNIFEM (1993a); Walker (1995).

Dairy products

Milk from many different sources (for example cows, goats, water buffalo, camels, horses and sheep) is important in many developing countries. The degree to which milk consumption and processing occurs differs from region to region. It is dependent upon many factors including geographic and climatic conditions, availability and cost of milk, food taboos and religious restrictions. Where processing does exist, many traditional techniques can be found for making indigenous milk products.

Fermented milk products such as yoghurt and soured milk contain bacteria from the *Lactobacilli* group. These bacteria aid digestion and help prevent illness caused by other bacteria. In addition, fermentation removes milk sugar (lactose) from milk and thus enables people that suffer from lactose intolerance to consume these dairy products.

Milk is a perishable commodity and spoils very easily. Its low acidity and high nutrient content make it the perfect breeding ground for bacteria, including those that cause food poisoning (Part I). Bacteria from the animal, utensils, hands and insects may contaminate the milk, and their destruction to preserve the milk and make it safe is the main reason for processing. Infections in the animal which cause illness may be passed directly to the consumer through milk. It is therefore extremely important that quality control tests are carried out to ensure that the bacterial activity in raw milk is of an acceptable

Milk comes from many different sources

level and that no harmful bacteria remain in the processed products. Dairy products should not be processed by inexperienced people, and training should be given to deal with the risks associated with these products.

Nutritional significance

Milk is regarded as nature's most complete food because it provides many of the nutrients which are essential for growth. It is an excellent source of protein and fat and has an abundance of vitamins and minerals, particularly calcium. Milks from different species differ in the levels of fat and protein they contain. For instance, buffalo and sheep milk are particularly rich in fat (see Table 15). This has implications for the type of processed products they are suitable for.

Table 15. Fat and protein content of selected milks

Species	*Fat (%)*	*Protein (%)*
Cow	3.7	3.5
Goat	4.25	3.52
Sheep	7.9	5.23
Indian buffalo	7.38	3.6
Camel	5.38	2.98
Horse	1.59	2.69
Llama	3.15	3.9

Processing

The flow chart on the opposite page indicates the various stages of processing raw milk into a number of products and the equipment requirements.

Types of processing

Liquid milk

Milk can be preserved if it is heated to destroy the bacteria or cooled to slow their growth. Cooled raw milk is widely available is some countries, and milk is also cooled prior to delivery to dairy companies. Simple milk coolers can be found in the Directory (sections 11.0 and 8.1). Pasteurization and boiling are the two most commonly used heat treatments. Boiling centres are very common in developing countries, where raw milk is boiled in bulk and then distributed daily in jerry cans. Technically, it is possible for pasteurization to be carried out on a small scale, but it is more usually performed on a larger scale because of the higher investment in equipment (Directory section 28.0) and the need for qualified and experienced staff, and accurate and strictly controlled hygienic processing conditions. 'Sterilized' milk is produced by two methods: heating bottled milk in a retort (similar to canning) and ultra-high temperature (UHT) processing. Neither is suitable for small-scale processing.

TESTING PROCEDURES

Milk fat

The price paid for milk is usually dependent upon the milk-fat content (Table 15), and this may be determined either at the collection stage or at the dairy using a hydrometer (Directory sections 11.0 and 38.2), which measures the specific gravity. More complex equipment that measures fat content, known as a butyrometer (sections 11.0 and 38.0), is used by larger processors as an aid to detect adulteration.

Bacterial activity

Two tests that are used to check the microbiological quality of raw milk use either methylene blue or resazurin dyes. These tests determine whether the milk is accepted or rejected. Both work on the principle of the time taken to change the colour of the dye, which is proportional to the number of micro-organisms present (the shorter the time taken, the higher the bacterial activity). It is preferable to use the resazurin test (see p. 118) as this is less time-consuming (Directory section 11.0 and 38.0).

Summary of equipment needed for dairy processing

Processing stage	*Equipment required*	*Directory section*
Cooling	Refrigerated storage	8.1
↓	Thermometer	38.5
	Bottle-cooling system	
Test for fat content	Butyrometer	11.0, 38.0
Test specific gravity	Hydrometer	11.0, 38.2
Test bacterial activity	Methylene blue or resazurin dyes	
↓	Phosphatase test	
	Thermometer	38.5
	Basic laboratory equipment	38.0, 11.0
Filter	Filter cloth	11.0
↓	Filter press	18.1
Homogenization ↓	Homogenizer	21.0
Pasteurization	Boiling pan	27.0
↓	Pasteurizer	28.0
	Heat source	25.0
	Thermometer	38.5
	Refrigerator	8.1
Separation of milk fat ↓	Cream separator/centrifuge	11.0, 11.1, 34.2
Churning ↓	Butter churns	11.1
Kneading/working ↓	Butter pats	
Ice cream making	Ice cream maker	22.0
↓	Freezer	8.2
Cheese making	Cheese vat	11.2
↓	Cutters	11.2
	Strainers	11.0
	Cheese press	11.2
	Curing racks	
Filling/sealing	Liquid-filling machine	17.1
	Capping machine	26.2
	Manual heat-sealer	26.1
	Wrapping machines	26.3

The resazurin test

This test is based on the colour change of a dye (resazurin) from blue to pink.

1. Take a sample of milk and shake it well.
2. Take a 10ml sample of the milk into a sterile test tube.
3. Add 1ml of sterile resazurin, shake well and place in a water bath at 37.5 °C for 10 minutes.
4. Compare the colour of the sample with that of the standard using a comparator (available with the resazurin dye). If the colour is less than 3.5 (on a standard scale) the sample should be rejected.

Resazurin dye and the comparator should be available from chemical suppliers.

Phosphatase test

It is possible to ensure that pasteurization has been adequately achieved by testing for the presence of the enzyme phosphatase, present in raw milk. The destruction of phosphatase shows that milk has been sufficiently heat-processed, because this enzyme is destroyed by pasteurization conditions.

Homogenization

Homogenization breaks up the oil droplets in milk and prevents the cream from separating out and forming a layer. Homogenizers are designed for industrial-scale production, but it is possible to purchase smaller versions, although even these are very expensive.

Pasteurization

Pasteurization is a relatively mild heat treatment (below 100 °C) which is used to extend the shelf life of milk for several days. It preserves the milk by the inactivation of enzymes and destruction of heat-sensitive micro-organisms, but causes minimal changes to the nutritive value or sensory characteristics of a food. Some heat-resistant bacteria survive to spoil the milk after a few days, but these bacteria do not cause food poisoning. The time and temperature combination needed to destroy 'target' micro-organisms varies according to a number of complex interrelated factors. For milk, the heating time and temperature are either 63 °C for 30 minutes or 72 °C for 15 seconds. Only the former combination is possible on a small scale, and for this the simplest equipment required is an open boiling pan. Better control is achieved using a steam jacketed pan (Directory section 27.1), and this can be fitted with a stirrer to improve the efficiency of heating. Both of these are batch processes which are suited to small-scale operation, but the quality of the milk is lower than that obtained using a plate heat exchanger (Directory section 28.0). In some countries, flavoured pasteurized milk has become a very popular product.

Sterilization

In-bottle sterilization is a more severe heat treatment designed to destroy all contaminating bacteria. The milk is sterilized at a temperature of 121 °C maintained for 15 to 20 minutes, using a retort or pressure cooker. Unlike pasteurization, this process causes substantial changes to the nutritional and sensory quality of the milk. However, sterilization is not recommended for small-scale production for the following reasons:

- the cost of a retort and ancillary equipment is very high
- it is essential that the correct heating conditions are carefully established and maintained for every batch of milk that is processed
- if the milk is not heated sufficiently, there is a risk that micro-organisms will survive and grow inside the bottle; in low acid foods such as milk, this can cause severe food poisoning
- because of the potential dangers from food poisoning, the skills of a qualified food technologist/microbiologist are required to routinely examine samples of sterilized milk that have been subjected to accelerated storage conditions; this requires a supply of microbiological media and equipment

- in summary, the process of sterilization requires a considerable capital investment, trained and experienced staff, regular maintenance of sophisticated equipment and a comparatively high operating expenditure.

FILLING/PACKAGING EQUIPMENT

At a small scale, milk and other dairy products are sold in polythene bags, which may be simply tied or heat-sealed. Glass bottles can be sealed with either foil or metal caps and plastic bottles are suitable for pasteurized milk when glass is not available or too expensive. Laminated cardboard cartons are too expensive for most small producers.

STORAGE

Pasteurized milk has a shelf life of two to three days if kept at around 4°C. Maintaining this low temperature causes a substantial increase to the cost of transportation and distribution, and has prevented the development of small-scale pasteurized milk businesses in many countries.

SEPARATION OF MILK COMPONENTS

When milk is left to stand for some time, fat globules rise to the surface to form a layer of cream. This can be separated, leaving behind skimmed milk as a second product. There are different types of cream each with different fat concentrations: single (or light) cream contains 18 per cent milk fat, whereas double (or heavy) cream normally contains 30 per cent milk fat. Cream is a luxury item that has little demand in many developing countries.

Separation can very simply be achieved by removing the cream with a spoon; however, this is a slow process during which the cream may spoil. For this reason it is more usual to use a manual or powered cream separator (or centrifuge) (Directory sections 11.1 and 34.2).

FERMENTATION

A wide variety of fermented dairy products are made using lactic acid bacteria (see Part I), including yoghurt, lactic butter and cheese. These are described below. The technology for producing cultured milk products is based upon the microbial conversion of the milk-sugar lactose to lactic acid (lactic acid accounts for the characteristic 'sourness' of such products). This may occur by allowing the milk to sour naturally, but it is better to introduce the lactic acid bacteria as a starter culture, in the form of a small quantity of previously cultured product or as a commercially prepared culture. Although most do not require sophisticated equipment for small-scale production, there are specific pieces of equipment that are used for each, such as cheese presses, cutters etc. (Directory section 11.2) and yoghurt incubators (section 11.3).

Dairy products

Soured milk

Traditional soured milk is a thick clotted milk which has a stronger flavour and a more acidic taste than yoghurt. It has a similar shelf life of three to eight days and is used as a drink or as an accompaniment to meals in some countries. Preservation is due to the production of lactic acid by naturally occurring bacteria in the untreated milk. The high levels of acid inhibit the growth of spoilage bacteria and pathogens that may be present in the raw milk. Raw milk is allowed to sour by naturally occurring bacteria. It is transferred into 5 litre gourds or other containers fitted with lids and placed in a warm ventilated room for one or two days. The main quality control point is the correct incubation temperature to allow rapid production of lactic acid by naturally occurring lactic acid bacteria. If the temperature is too high the bacteria will be destroyed; if it is too low there may be insufficient acid production to form the product.

The product may be used in portions directly from the culture vessel, or alternatively the milk may be poured into pots and allowed to ferment in them. The product has a short shelf life and only requires protection against dust, insects etc. It should be stored in a cool place away from sunlight. Clay pots, gourds and wooden or ceramic bowls are each used traditionally and are suitable provided proper hygiene is observed in their preparation and cleaning, particularly if they are to be reused.

Wagashie

In West Africa a dairy product is prepared by souring, cooking and frying cow's milk. It is hard, has a slightly sour and fermented flavour and may be a natural white or coloured dark red. It is used instead of meat in soups and stews. The principles of preservation are as follows:

- to acidify the milk by fermentation
- to destroy enzymes and spoilage micro-organisms in the milk by boiling the curdled milk
- to reduce the moisture content and destroy spoilage micro-organisms by frying the cooked and hardened milk.

Milk is poured into a tray and placed in the sun for two to three hours. Heat from the sun curdles the milk. It is then boiled for one hour. The boiling mass is stirred during cooking until it thickens and hardens. *Karandafi* (dark red dried leaves from the sorghum plant) may be added to the boiling mass to impart a red colour to the product. It is then cooled, moulded into small round pieces and fried in coconut oil until slightly browned. The product is not elaborately packaged and is simply displayed on open trays or in glass boxes. The shelf life may range from one to five months, although it is usually sold more quickly. Over time the product becomes sour with a fermented flavour.

Lactic butter

Whole fresh milk is pasteurized and then cooled to 37 °C before adding a starter culture (this may be a small amount left over from a previous batch). The milk is then incubated for several hours (or overnight) in a warm place (30–37 °C) until set. The resulting yoghurt is cooled for several hours, and then salted and churned to produce the lactic butter.

Traditional butter

Cow's milk is put into well-cleaned and smoked clay containers, then left to solidify for three to four days depending on the ambient temperature. It is then agitated by shaking the tightly closed container (with a small vent for releasing the air that builds up during the first few minutes of shaking). The butter granules float on the surface of the buttermilk, growing in size as the agitation nears completion. The butter curds are collected by hand and washed with clean water two or three times. The butter is packed in aluminium, clay or plastic containers or in strong plastic bags for transportation.

Butter

Butter is a semi-solid that contains approximately 80–85 per cent milk fat and 15–20 per cent water. It is yellow or white in colour, with a bland flavour and a slightly salty taste. It is a valuable product that has a high demand for domestic use in some countries and as an ingredient in other food processing (e.g. for confectionery and bakery uses). The principles of preservation are:

- to destroy enzymes and micro-organisms by pasteurizing the milk
- to prevent microbial growth during storage by reducing the water content, by storing the product at a low temperature, and optionally by adding a small amount of salt during processing.

Cream is churned to disrupt the fat and water emulsion and as a result the milk fat separates out into granules. Buttermilk is drained off and can be used as either a beverage or as an ingredient in animal feed. Clean water equivalent in weight to the buttermilk is added to the churn to wash the granules, and churning is continued for a short time to compact the butter. It is then removed from the churn and the butter is kneaded to a smooth and pliable texture using butter pats (Directory section 11.1). Alternatively for higher production rates a specially designed kneader (section 11.1) can be used. It is then formed into blocks with the butter pats and packed in either grease-proof paper or foil wrappers. Because of its high fat content, butter must be stored below 10 °C to prevent rancidity and undesirable off flavours. The water droplets in butter can also allow bacteria to grow if it is not kept under cool conditions.

Ghee

Ghee is a clear golden brown fat (below 30 °C) or oil (above 30 °C) with a characteristic flavour of milk fat. It is made from cow's or buffalo milk and has a high demand in Asia and parts of Africa as a cooking oil for domestic use and as an ingredient for local food products (for example bakeries, confectionery manufacturers). It is preserved by heat that destroys enzymes and contaminating micro-organisms and by the low moisture content which prevents microbial growth during storage. The main quality factors for ghee are colour, clarity, flavour and odour. Correct colour and clarity are mainly due to proper filtering. The taste, colour and odour are determined by the time and temperature of heating. Overheating produces a burnt taste and odour and a dark colour. Ghee has a long shelf life if stored in an air-tight, light-proof and moisture-proof container and a cool storage room to slow down rancidity. Metal containers are normally used. They should be thoroughly cleaned, especially if they are reusable, and they should be made airtight. Alternatives include coloured glass jars with metal lids, or ceramic pots sealed with cork/plastic stoppers. Ghee is usually stored at room temperatures as cold storage affects the granular texture. Thus ghee is useful for those consumers with no access to refrigeration.

Yoghurt or curd

Yoghurt is made from a variety of milks, whereas the term 'curd' is usually used for fermented and thickened buffalo milk. Both products have a smooth, firm, white gel with a characteristic acidic taste, with a creamy consistency. They are eaten in Asia with sugar or treacle as a dessert, mixed with rice or as a component of a curry meal. Stirred yoghurt can be fermented in the mixing container, and to make set yoghurt the inoculated milk should be poured into the individual pots before fermentation. Other similar products can be made from goat's or mare's milk. Yoghurt is used in Africa, Asia and Latin America to accompany a main meal, as a dessert or as a dressing for vegetable salads. It has a shelf life of three to eight days depending on the storage conditions. The principles of preservation for yoghurt are:

- pasteurization of the raw milk to destroy contaminating micro-organisms and enzymes
- an increase in acidity due to the production of lactic acid from lactose; this inhibits the growth of food-poisoning bacteria
- storage at a low temperature to inhibit the growth of micro-organisms.

Milk is heated to 70 °C and held at this temperature for 20–30 minutes. It is then cooled to 44–42 °C by dipping the container into a tub containing cold water. The cooled milk is inoculated with selected strains of actively growing starter cultures of *Lactobacillus bulgaricus* and *Streptococcus thermophilus* (either bought as a dry powder or from a portion of a previous batch). A temperature of 42–44 °C is maintained for approximately five hours until the desired degree of acidity is achieved to form the correct consistency. It is then cooled and stored in a cool place until next morning to check for curd formation. The main hygienic requirements are:

- to thoroughly clean and sterilize all equipment and utensils with chlorine solution or boiling water before and after processing
- to filter milk after milking to remove visible dirt and 'ropiness'

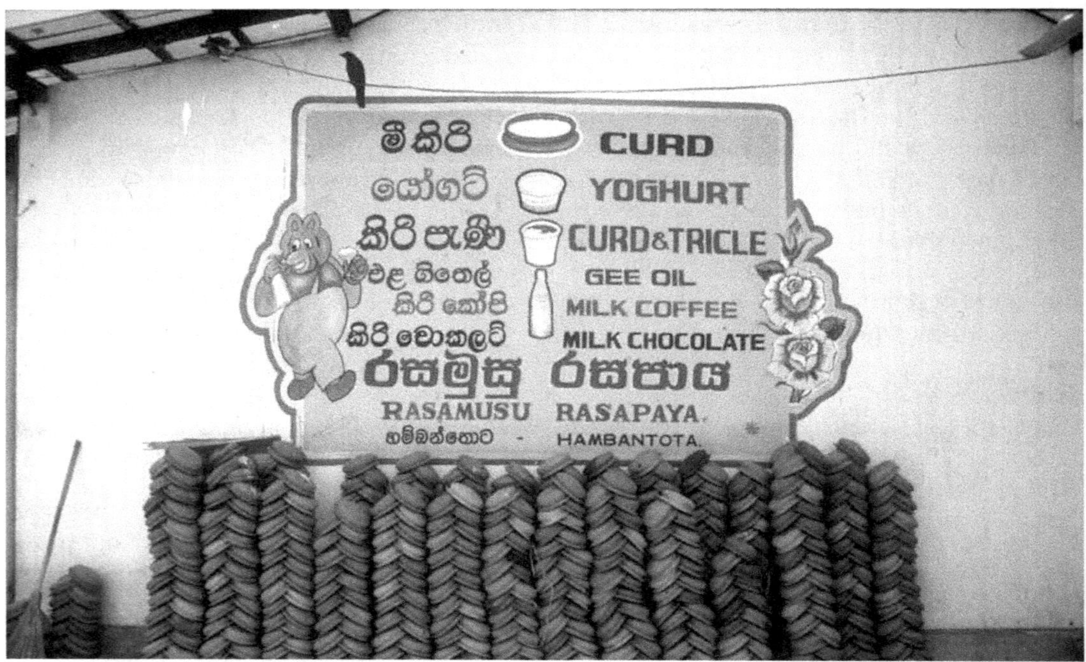

Selling curd in traditional clay pots

- to cool milk immediately to control further growth of micro-organisms and enzyme activity.

Milk which is likely to contain antibodies should not be used as the antibodies will inhibit the action of lactic bacteria.

The micro-organisms that produce yoghurt are most active within a temperature range of 32–47 °C, and ambient temperatures are therefore not high enough in many countries. Small commercially available yoghurt-makers consist of an electrically heated base and a set of plastic or glass containers. There are other simple and inexpensive ways of incubating yoghurt such as an insulated box containing a light bulb as a heater, keeping the jars/pots surrounded by warm water, or by using thermos flasks (the latter is only suitable for stirred yoghurt) (Directory section 11.3).

To produce curd, buffalo milk is filtered and boiled, the scum is removed and it is cooled to room temperature. A few spoonfuls of a previous batch of curd are added and it is then well mixed and poured into clay pots. These are sealed by wrapping a piece of paper over the pot. The mixture is allowed to stand for 12 hours. The fermentation develops the characteristic flavour of the product. Curd is eaten or distributed and sold immediately after manufacture, as storing curd for more than a few days at room temperature makes it too sour.

The short shelf life of yoghurt does not require sophisticated packaging, and the product only requires protection against dust, insects etc. Pots should be stored in a cool place away from sunlight and preferably in a refrigerator. Clay pots, gourds and wooden or ceramic bowls are each used traditionally and are suitable provided proper hygiene is observed in their preparation and cleaning, particularly if they are to be reused.

Traditional cheese

Unfermented cheese is white in appearance and similar to European cottage cheese, but more sour in taste, and with a pronounced cheese flavour. It is used as a component of traditional dishes and is made by coagulating the milk solids in butter milk. Fermented butter milk is heated gently in a clean pot until coagulation is complete after about 30 minutes. The cheese is collected by draining off the whey and putting the curds on a straw screen to drain. The product must be sold or eaten within 24 hours if refrigeration is not available. The main quality control points are the temperature and time of heating, which determine the product quality and yield: overheating results in loss of quality and off flavours, whereas under-heating lowers the yield and may risk passing on food-poisoning bacteria to consumers.

Cheese

Cheese is made from milk by the combined action of lactic acid bacteria and the enzyme rennet. Just as cream is a concentrated form of milk fat, cheese is a concentrated form of milk protein. The differences in cheeses that are produced in different regions result from variations in the composition and type of milk, variations in the process and the bacteria used. The different cheese varieties can be classified as either hard or soft: hard cheeses such as Cheddar and Edam have most of the whey drained out and are pressed. Soft cheeses such as paneer contain some of the whey and are not pressed. The principal steps of a cheese-making process are:

- raw milk is pasteurized to destroy most enzymes and contaminating bacteria
- fermentation by lactic acid bacteria increases the acidity which inhibits the growth of food-poisoning and spoilage bacteria
- the moisture content is lowered and salt is added to inhibit bacterial and mould growth.

Starter culture is added to the milk at approximately 2 per cent of the weight of milk in a fermentation vessel made from either aluminium or stainless steel. The rennet can be bought from specialist suppliers and should be added at 1 per cent of the weight of milk. It alters the milk proteins and allows them to form the characteristic curd. The milk is allowed to stand until it sets to a firm curd, which is then cut into cubes to assist the removal of whey. The curd is then cooked at 40 °C for 20 minutes to firm it. After cooling, the whey is drained off and the curd is cut into small pieces and pressed to ensure that most of the whey has been removed. The pieces are placed in cheese moulds and pressed with weights. Ripening in a room where the temperature and humidity are controlled allows the development of gas in some cheeses and the development of flavour. The longer the ripening process the stronger the flavour.

The packaging requirements differ according to the type of cheese produced. Hard cheese has an outer protective rind which protects it from air, micro-organisms, light, moisture loss or pick-up, and odour pick-up. Cheese should be allowed to 'breathe', otherwise it will sweat. Suitable wrapping materials are therefore cheesecloth or greaseproof paper. Cheese should be stored at a relatively low temperature (4–10 °C) to achieve a shelf life of several weeks/months. Soft cheeses are stored in pots or other containers, often in brine, to help increase their shelf life of several days/weeks.

Ice cream

Ice cream is a frozen mixture which contains milk, sugar, fat, and optional thickeners (e.g. gelatin), colouring and flavouring. It may be sweetened and flavoured in numerous ways with nuts, fruit pieces, and natural or artificial flavours and colours. The principles of preservation are:

- pasteurization to destroy most micro-organisms and enzymes
- freezing to inhibit microbial growth.

Pasteurization is carried out by heating to 65 °C for a period of 30 minutes. Ice cream is a complex mix of small ice crystals and air bubbles in a milk-fat/water emulsion. To achieve this, it is necessary to cool the mixture quickly to produce small ice crystals and at the same time incorporate air into the product by beating. Ice cream makers (Directory section 22.0) are available commercially and work on the following principle: the mixture is placed in a bowl which is kept at a low temperature (either surrounded by ice and salt, or having been chilled in a freezer). It is then agitated by a manually operated rotor or by a powered stirrer. At the end of this process the ice cream should be at a temperature of approximately minus 5 °C, and be partly frozen. It is then either sold directly or hardened in a freezer at minus 18 °C. Ice cream may be transported in an insulated box (e.g. for sale from a bicycle). It is especially important to guard against thawing and refreezing as this will cause changes in texture and mouth-feel, and there is the increased possibility of food poisoning by contaminating food-poisoning micro-organisms.

Milk confectionery

Milk-based sweets are common throughout the Indian subcontinent, where there are thousands of sweet or *mishti* makers. A variety of sweets are made using *channa* as the basic ingredient. *Channa* is made by boiling milk and separating the whey to form a solid material. Other ingredients, including sugar, coconut, almonds, flavourings and colourings, are added to the *channa* to produce a range of sweets. *Sondesh* is one such typical product.

Sondesh

Ingredients per 1 kg *channa* include 300–500 g sugar, cardamom (optional), food colouring (turmeric is widely used), flavouring (optional). The production process involves mixing and kneading the sugar with the *channa* until it becomes smooth, and heating slowly with constant stirring until the mixture becomes sticky. Portions taken from it at this stage can easily be formed into a ball. Flavouring (such as cardamom) may also be added at this stage. The mixture is spread onto a tray coated with ghee or oil and allowed to cool and set. It is then cut or moulded into the desired size and shape.

Khova

Khova is a product obtained by evaporation of milk. It is a semi-solid compact mass, creamish-white in colour, granular in texture, with a milky fatty flavour. It is used in the preparation of many sweetmeats (*gulab jamun*, *burfi*, milk cake, *doodh peda* etc.). It has a shelf life of two to three days under refrigerated conditions.

Fresh buffalo milk (with a minimum of 5 per cent fat) is boiled over a high heat with constant stirring until the milk starts to coagulate, which is marked by an abrupt change in colour. It is then heated at a low heat with regular stirring to scrape the milk solids from the sides of the pan. The end point is marked by the solid mass leaving the sides and bottom of the pan. It is then pressed to form a compact mass and stored for two to three days under refrigerated conditions.

Gulab jamun

Gulab jamun is a popular milk-based sweet made with *khova*, refined flour and cane sugar. It is round or elliptical in shape with a deep brown, slightly crisp, outer surface and a dull white, soft and porous inside. Always kept floating in sugar syrup, the product has a distinct flavour of deep fried milk solids, sugar syrup and added flavours.

Gulab jamun

Khova is broken to loosen the mass, and flour and baking powder are mixed in (900 g *khova*, 105 g refined flour, 7.5 g baking powder). It is then kneaded to form a smooth and soft dough. A sugar syrup is prepared by dissolving 3 kg of sugar in an equal quantity of water and boiled for five minutes. If desired, essence can be added. Small quantities of dough are shaped into balls of 2 cm diameter or oval forms and fried until they turn deep brown. They are immediately transferred to the sugar syrup, where they swell in size and become soft as they absorb the syrup. They are stored in glass jars.

Shrikhand

This product is a semi-solid smooth paste of creamish-white colour which has a sweetish sour taste. It is normally consumed as a special dish or as a dessert after meals. Acids produced during fermentation increase the acidity. Coagulation of milk removes part of the moisture and addition of sugar further increases the solids content to reduce the chances of microbial spoilage.

Pasteurized milk is cooled to room temperature (20–30 °C), inoculated with a previously set curd and left for about 10 to 12 hours for coagulation. The curd is broken and placed in a muslin cloth, and the whey is drained for 5 to 6 hours. The mixed mass of curd and sugar (500 g per litre of milk) is beaten until a smooth soft paste is formed, packed in plastic tubs and stored in a refrigerator.

Further reading

See: Ash (1983); Biss (1988); J. Dubach (1989); S. Dubach (1992); FAO (1990); Rothwell (1985); UNIFEM (1996).

DIRECTORY

Equipment directory

1.0 Bee-keeping equipment

Bee-keeping is not technically 'food processing', but this activity is appropriate for small-scale processors and producers, and is essential for the production of high quality honey. A range of bee-keeping and honey-processing equipment is therefore included in the directory.

BEE-KEEPING EQUIPMENT
Honey bee foundation and all bee-keeping and honey-processing equipment, such as hive components, protective clothing, honey pumps, bottling machines etc.
Bee Keeper's Supermarket, South Africa

HONEY-MAKING EQUIPMENT
Machinery etc. suitable for bee-keeping and honey processing.
Graze, Germany

BEE-KEEPING EQUIPMENT
Equipment for bee-keeping and honey processing.
Highveld Honey Farms, South Africa

BEE-KEEPING AND HONEY-PROCESSING EQUIPMENT
Full range of bee-keeping, queen-rearing and honey-processing equipment.
Steele & Brodie Ltd, UK

HIVE COMPONENTS
Manufactured hive components in selected pine.
Price code: 1
Stilfontein Apiaries, South Africa

STANDARD MACHINE FOUNDATION EMBOSSER
Creates a foundation for the bees to build their honeycomb onto.
Power source: electric/manual
Price code: 3
Tom Industries, USA

MIGHTY MINI
Machine that creates a foundation for bees to build their honeycomb onto.
Power source: manual
Price code: 2
Tom Industries, USA

SMOOTHING ROLLERS
Machine that creates a foundation for bees to build their honeycomb onto.
Power source: electric
Price code: 3
Tom Industries, USA

SHEETING MACHINE
Machine that creates a foundation for bees to build their honeycomb onto.
Price code: 4
Tom Industries, USA

BEEHIVES
Quality hives and frames.
Price code: 1
Woodlands Beehives, South Africa

Hive components are also manufactured by Beeswax Comb Foundation, South Africa.

1.1 Honey-processing equipment

HONEY EXTRACTOR
A centrifuge for the efficient extraction of honey from frames of uncapped honey. Consists of a stainless steel cylindrical container with centrally mounted radial cage to support frames of uncapped honey.
Power source: electric
Price code: 4
Alvan Blanch, UK

HONEY EXTRACTOR
A honey centrifuge for the extraction of honey from uncapped honeycombs.
Power source: electric/manual
Price code: 2
Swienty A/S, Denmark

ELECTRIC UNCAPPING PLANE
A powerful electric heating element with a cutting edge of 10 cm. Used for honey to keep the honeycomb square.
Power source: electric
Price code: 1
Swienty A/S, Denmark

DANA API MANA
A reliable and accurate machine for filling jars with honey, with a cut-off mechanism to prevent an extra drip. Capacity: 300–400 jars/hour.
Power source: manual
Price code: 2
Swienty A/S, Denmark

FILLING MACHINES
Suitable for filling jars using an electric motor or compressed air.
Power source: electric
Price code: 3
Swienty A/S, Denmark

CONICAL STRAINER
Strainer with nylon mesh. Very simple and very effective with honey and other liquids.
Power source: manual
Price code: 1
Swienty A/S, Denmark

STIRRING SCREW
Used for honey processing.
Power source: manual
Price code: 1
Swienty A/S, Denmark

WAX FOUNDATION MACHINE
Hand driven with plane roller and foundation roller.
Power source: manual
Price code: 3
Swienty A/S, Denmark

2.0 Blanchers

Blanching is an important processing stage for a number of reasons: primarily it inactivates enzymes that cause deterioration in colour and flavour during drying and subsequent storage of vegetables and some fruit. Blanching may be carried out using water or steam.

SOYBEAN BLANCHING UNIT
This machine imparts wet heat treatment to soybeans to eliminate anti-nutritional elements. It is designed to use agricultural waste as fuel. It can also be used for the blanching of potato slices. Suitable for home- to cottage-level processing units. Capacity: 20 kg/hour.
Power source: biomass
Central Institute of Agricultural Engineering, India

DEPAR 250 DRY BLANCHER
This blancher is used for removing the layer of red skin which remains on the groundnut after roasting and to separate the nut kernels. Capacity: 200–1000 kg/hour.
Gauthier, France

BLANCHER 350
Capacity: 0.5–5 tonnes/hour.
Gauthier, France

2.1 Steam blanchers

BLANCHER
This machine has a hot water tank and a perforated basket with sloping base that incorporates a swing off system with a cooling tank. It is all fitted on an aluminium stand, steam heated with contacting parts of aluminium.
Power source: electric
Price code: 1
Gardners Corporation, India

FOOD STEAMING/BLANCHING MACHINE
Capacity: 100 kg/hour
Price code: 2
Sahathai Factory, Thailand

3.0 Bottle washing equipment

A variety of bottle washers are available. They include the following variations:

- simple manual bristle washer
- hydro washer (cleaning is performed by jets of water)
- soaker washer (bottles are completely immersed)
- combination of the above.

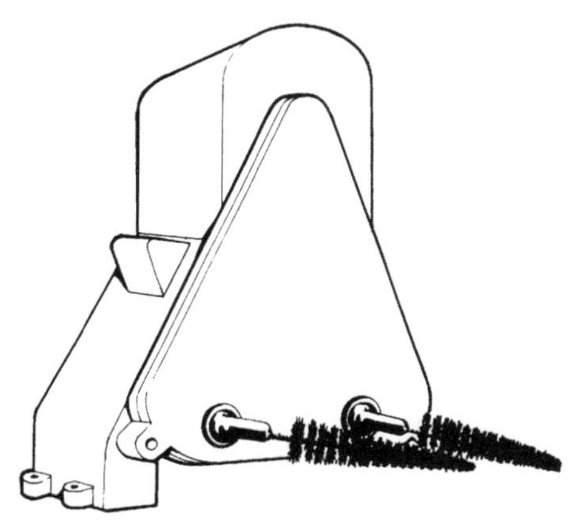

BOTTLE WASHING, FILLING, CAPPING/SEALING MACHINE
This combination machine can operate automatically or semi-automatically and is suitable for a range of bottles from 0.5 to 1.5 litre volume. Washing operation is intermittent. Filling and capping/sealing operation is continuous. Low percentage of defective products. Capacity: 1000–2000 bottles/hour.
Power source: electric
Price code: 4
ALFA Technology Transfer Centre, Vietnam

BOTTLE WASHER
This machine comes with two- and four-head washers.
Eastend Engineering Company, India

BOTTLE WASHING MACHINE
This machine has two brushes and a drive motor.
Power source: electric
Price code: 1
Gardners Corporation, India

GLASS BOTTLE AND JAR DRYING CABINET
This machine has a capacity of 340 jars of 500 g, and takes 25–30 minutes per batch. There is a mechanism for vapour escape. It is heated by two air heaters.
Power source: electric
Price code: 1
Gardners Corporation, India

ROTARY BOTTLE WASHING MACHINE
Capacity: 4000 bottles/hour
Power source: electric
Mark Industries (Pvt) Ltd, Bangladesh

BOTTLE WASHING MACHINE
Capacity: 1000+/hour
Pharmaco Machines, India

BOTTLE DRYER
This machine is complete with dial thermometer, thermostat, heater, motor, and cross-flow type circulation with a fan fitted on the side.
Power source: electric
Rank and Company, India

BOTTLE WASHING MACHINE
This machine has two brushing heads with rinsers and a soaking tank. Capacity: 5000 bottles per day.
Power source: electric
Rank and Company, India

Bottle washing equipment is also manufactured by Geeta Food Engineering, Narangs Corporation and Techno Equipments, India.

4.0 Bread-making equipment

A range of bread-making equipment is available, from dough mixers and kneaders through to moulds, dough cutters, baking pans, and slicing and wrapping equipment. Complete bread-making units are available for a range of breads, biscuits and confectionery items and for specialized breads. See also sections 24.0 (mixers), 25.0 (ovens) and 25.2 for integrated baking machines for *idli*, *dosa* and chapattis.

BISCUIT, BREAD AND CONFECTIONERY MAKING PLANT
Machines for the complete process of biscuit, bread and confectionery making.
Power source: electric
M/S Mangal Engineering Works, India

4.1 Dough mixers

DOUGH MIXING AND BEATING MACHINES
Machines with grinder attachments.
Power source: electric
AMI Engineering, India

DOUGH MIXER
Two speed, with a 2.4/3.6 hp motor.
Capacity: 25 kg
Power source: electric
ANLIN, Peru

BAKERY MACHINERY
All types of bakery machinery, including mixers, kneaders, dough sheeters etc.
Capacity: 50–300 kg/hour
Power source: diesel/gas/electric
Apple, India

DOUGH KNEADER
Designed to mix flour with other ingredients. Supplied with one mixing pan mounted on a wheeled trolley enabling free movement from one place to another. Motor 3 hp. Capacity: 50–170 kg
Power source: diesel/electric
Price code: 4
Baker Enterprises, India

PLANETARY MIXER
Machine designed to suit large-, medium- and small-scale bakeries for uniform mixing of cake batters and small quantities of bread and pie doughs. Attachable items: three stainless steel beaters and a stainless steel pan. Three speeds are available for kneading, mixing or whipping. Capacity: 25–30 litres
Power source: diesel/electric
Price code: 4
Baker Enterprises, India

FLOUR KNEADING MACHINE
Price code: 2
Gardners Corporation, India

DOUGH MIXER
Capacity: 25 kg/hour
Power source: electric
Industria Peruana Comercializadora Techno Pan Equipamiento Integral, Peru

FLOUR DOUGH MIXING MACHINE
Capacity: 50 kg/hour
Power source: electric
Price code: 2
Krungthep Chanya, Thailand

A flour kneader is also manufactured by Technoheat Ovens and Furnaces Pvt Ltd, India.

4.2 Dough and pastry moulds

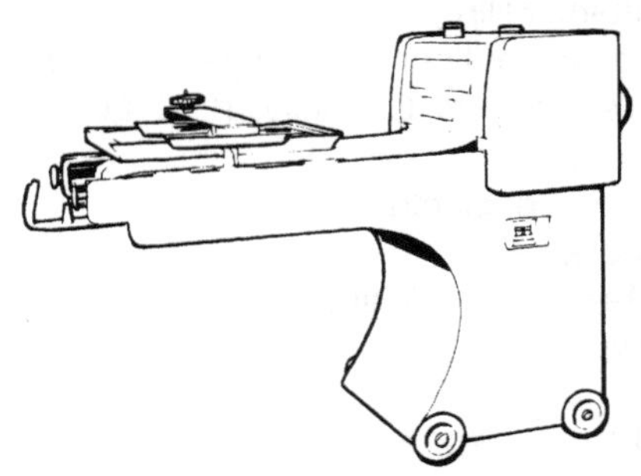

DOUGH DIVIDERS
Designed to scale dough into pieces of equal weight which can be adjusted by a setting device.
Power source: diesel/electric
Price code: 4
Baker Enterprises, India

MOULDER
Designed for continuous and uninterrupted operation at high speed and for perfect moulding action. Three-stage sheeting is accomplished by effective use of perfectly machined and precisely adjustable stainless steel rolls.
Power source: diesel/electric
Baker Enterprises, India

DOUGH (MAIDA) KNEADING MACHINE
Tank covered with stainless steel. Main shaft and handles are also stainless steel.
Capacity: 50–100 kg/hour
Power source: electric
Bijoy Engineers, India

BREAD MOULDING MACHINE
Capacity: 500–700 loaves/hour
Power source: electric
Price code: 2
Gardners Corporation, India

POWER OPERATED DOUGH DIVIDER
Power source: electric
Price code: 2
Gardners Corporation, India

MANUAL DOUGH DIVIDER
Power source: manual
Industria Peruana Comercializadora Techno Pan Equipamiento Integral, Peru
Baker Enterprises, India

A dough moulder/divider is also manufactured by Alven Foodpro Systems (P) Ltd, India.

4.3 Bread slicers

SLICERS
Power and gravity feed slicers to suit the requirements of small- and large-scale bakeries. Capacity: 200–800 loaves/hour
Power source: diesel/electric
Price code: 4

BREAD SLICING MACHINE
Power source: electric
Bijoy Engineers, India

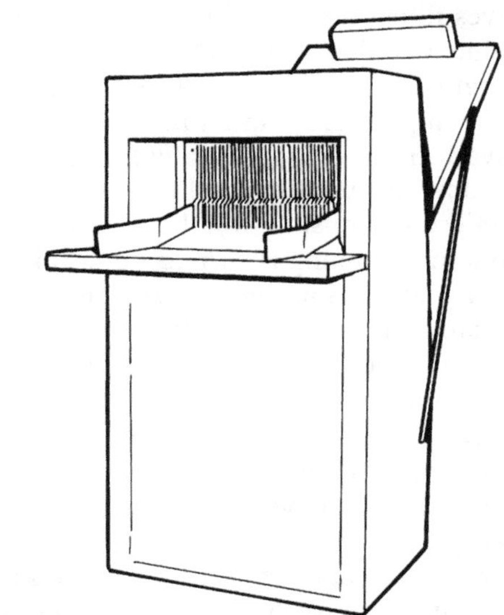

BREAD SLICING MACHINE
Capable of handling bread up to 25 cm long.
Power source: electric
Price code: 2
Gardners Corporation, India

Bread slicers are also manufactured by Alven Foodpro Systems (P) Ltd, India, and Technoheat Ovens and Furnaces Pvt Ltd, India.

4.4 Bread wrapper

ELECTRIC BREAD WRAPPER SEALING MACHINE
Complete with thermostatic switch.
Power source: electric
Price code: 1
Gardners Corporation, India

5.0 Canning equipment

Perishable foods which are to be stabilized for extended storage must be sterilized to retain desirable quality factors (flavour, colour, aroma); to destroy enzyme activity and, most importantly, to kill micro-organisms which cause undesirable changes in the food quality and others which produce deadly toxins, such as *Clostridium botulinum*. Canning is perhaps the most widely used sterilization process. The process involves filling the can with food, fitting the lid using a double seam and heating the can in a retort to sterilize the food. Different retort temperatures and processing times (derived from complex calculations and equations) are used depending on the type of micro-organism to be destroyed. Commercial canning is carried out as either a batch or continuous process. For canning at the small scale, a batch system using retorts or pressure cookers is more appropriate.

Although canning is possible at a small scale, it is not a technology that is recommended, because of several constraints. For example, the equipment, material and operating costs are high. As well as the retort and canning equipment, the process requires other equipment, such as a steam generator and an air compressor. In addition, canning requires experienced workers to operate the equipment, maintenance engineers to service the equipment and microbiologists (and specialized laboratories and equipment) for examination of the products. It is a technology that should only be undertaken by experienced processors, especially if low acid foods such as fish, vegetables or meats are to be canned.

EASY OPEN ENDS
This equipment can be used on all types of consumer cans in tin or aluminium.
Ajay & Abhay (P) Ltd, India

VETERINARY/CATTLE & CAN WASHING EQUIPMENTS & UTENSILS
A range of canning equipment, including can-steaming blocks, can-washing troughs and mini-pasteurizers.
Dairy Udyog, India

CAN CLOSING MACHINE
This hand-operated can sealer is flywheel operated and has an automatic sealing operation capable of sealing cans up to 10 cm in diameter.
Power source: manual
Narangs Corporation, India

HOT LIFTING TONGS
Used for lifting hot cans of all sizes.
Narangs Corporation, India

Canning equipment is also manufactured by Cantech Machines, India, and Narangs Corporation, India.

5.1 Autoclaves/retorts/sterilizers

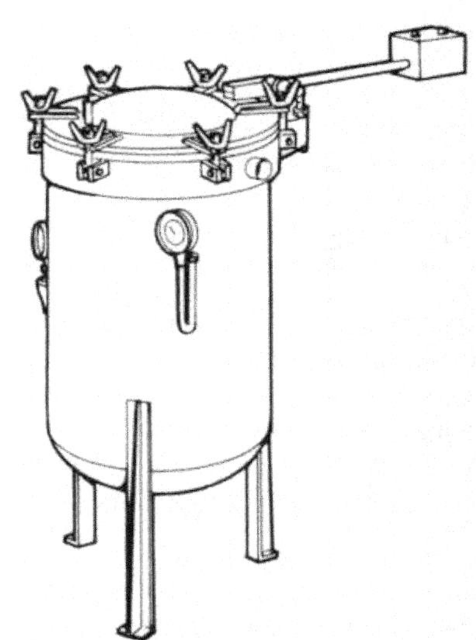

STERILIZING EQUIPMENT
Sterilizes food by blowing a flow of oxygen through a high voltage electric field to produce ozone.
Power source: electric
Price code: 3
Physics Institute of Ho Chi Minh City (Applied Electronics Division), Vietnam.

CAN STERILIZING MACHINE
Power source: water
Price code: 3
Sahathai Factory, Thailand

5.2 Steamers

STEAM BOILER
Provides a supply of steam for various processing operations. It incorporates a flame failure safety control. A set of pipework takes steam from the boiler to an inlet manifold on the steaming tanks – with pressure controls, gauges etc. Capacity: 96–960 kg/hour
Price code: 4
Alvan Blanch, UK

6.0 Centrifuges

Centrifuges are used to separate out the different components of food, for example, the cream from liquid milk, honey from honeycombs, fruit juice from pulp. By spinning the whole food at a high speed, the heavier components separate out and sink to the bottom.

STAINLESS STEEL CENTRIFUGAL PUMP
This centrifuge is used for a range of liquid foods. Capacity: up to 350 cubic metres/hour
Power source: electric
Fullwood Ltd, UK

CENTIFRUGE JUICER
Used for making fruit juice.
Narangs Corporation, India

Centrifuges are also manufactured by Premium Engineers Pvt Ltd, India.

6.1 Honey centrifuges

Honey centrifuges operate by forcing honey out of the combs. The extractor comprises a cylindrical container with a centrally mounted fitting to support the combs or frames of uncapped honey. There is also a mechanism to spin the frames which throws the honey out of the uncapped combs to the inner wall of the bowl. The honey then collects in the bottom of the bowl. A honey gate allows the honey to be drained out when required. Electrically operated high-speed extractors increase the speed gradually to prevent damage to the combs.

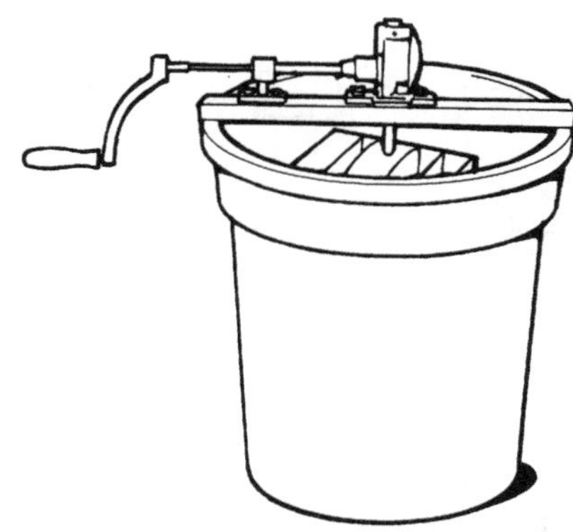

HONEY EXTRACTOR
A stainless steel cylindrical container with a centrally mounted radial cage to support frames of uncapped honey. This machine is used for the efficient centrifuging of honey out of frames of uncapped honey.
Power source: electric
Price code: 4
Alvan Blanch, UK

HONEY EXTRACTOR
Power source: electric/manual
Price code: 2
Swienty A/S, Denmark

6.2 Filter centrifuges

The centrifuge illustrated is an example of a manual filter centrifuge.

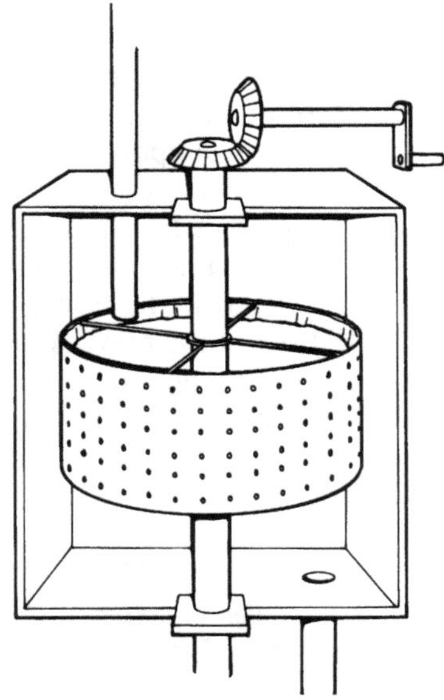

This can be made by taking a stainless steel drum with perforated holes (the drum of an old washing machine is perfect) and attaching it to a central shaft. The shaft is made to rotate through the use of toothed cogs and a flywheel. The drum is encased in an outer container. This can be made of any material, but if the filtrate is to be collected, the container must be made from food-grade plastic or a similar material. At the top of the outer container there is an inlet food chute for pouring in the liquid and at the base there is a collecting chute.

Mode of operation: the liquid is fed in at the top through the inlet chute. Simultaneously, the handle is turned and the drum made to rotate. This rotation causes the material to be thrown against the drum wall, and the filtrate drains out through the perforated holes and is finally collected at the base. If necessary, the drum can be lined with a filter cloth.

7.0 Cleaning equipment

Cleaning the raw material is the first stage of most processing operations. Depending upon the type of material used, this varies from simple washing to remove dirt to the removal of stones and other foreign matter through a combination of sieving, shaking and aspiration.

DESTONER

Used for removing stones from agricultural produce received from the fields. Capacity: 0.5–3 tonnes per hour
Power source: electric
Price code: 2–3
Acufil Machines, India

DESTONERS

Used for the removal of stones, metal etc. accurately and reliably from dry, granular and reasonably consistently shaped particles. The product flows through a regulated feed onto the vibrating bed where it is stratified by the combined action of the bed movement and the pressurized airflow from beneath the bed. Lighter material rises to the top, while heavier particles sink. Capacity: 750–8000 kg/hour.
Power source: electric
Alvan Blanch, UK

PRECLEANERS/GRADERS SAF/8

Ideal for pre-cleaning or commercial cleaning on a small scale.
Power source: electric
Alvan Blanch, UK

PRECLEANERS/GRADERS SA RANGE

These cleaners have a single air aspiration system – at the feed end – which makes them unsuitable for seed processing but ideal for conventional pre-cleaning duties.
Power source: electric
Alvan Blanch, UK

DESTONER
Efficiently removes stones, metal, glass and other materials heavier than seeds. Machines are available as vacuum types – fully sealed system suitable for all materials – and pressure types – open deck, suitable for low throughput, dust-free materials. Capacity: 100–1000 kg/hour
Power source: diesel/electric
Price code: 3
Forsberg Agritech (India) Pvt Ltd, India

STONE REMOVER – EPR 1000
Removes stones and other heavy materials from nuts and beans. Capacity: 1–4 tonnes/hour
Power source: electric
Gauthier, France

VACUUM DESTONER/PRESSURE DESTONER
The combined action of vibrating motion and fluidized bed creates separation of stones or any other impurities heavier than the product. Offers good air distribution and dust-free working through a totally enclosed aspiration system with cyclone. Controls of feed, air, deck elevation, pitch and speed facilitate almost complete removal of heavy impurities from products in one pass only.
Power source: electric
Price code: 3
Goldin (India) Equipment (Pvt) Ltd, India

DESTONER
3.6 hp motor/1700 rpm. Capacity: 350–400 kg/hour
Power source: electric
Price code: 3
Industrias Technologicas Dinamicas SA, Peru

HI-CAP DESTONER
For separating dry granular shapes that are similar in size and shape but different in weight.
John Fowler (India) Ltd, India

De-stoners are also manufactured by Sri Murugan Industries, India.

7.1 Fruit and vegetable cleaners

Fruit and vegetables generally require washing to remove dirt and traces of pesticide. The washing process should be delicate as these commodities bruise and spoil very easily.

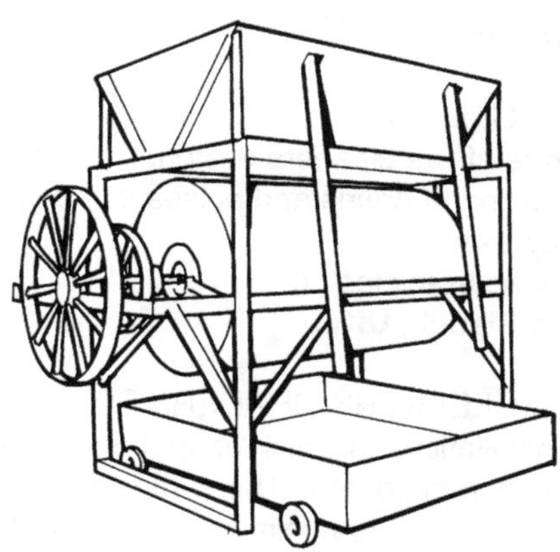

FRUIT WASHER (SPRAY WASHER)
This machine is used for the efficient washing of fruits prior to processing.
Power source: electric
Price code: 4
Alvan Blanch, UK

LV 2000 VEGETABLE WASHER
This machine gently washes salads, fruits and fish. It comes complete with a centrifugal dryer for salads. It will also defrost fish.
Capacity: 3 kg leafy vegetables per batch
Price code: 4
Crypto Peerless Ltd, UK

7.2 Grain cleaners

Grain cleaners are designed for use after winnowing primarily to separate all chaff, straw, weed seeds, broken and inferior seeds, dust and other contaminants. Cleaning and grading the grain make it more suitable for a number of uses: grain required for seed purposes should be clean and uniform; better prices can be obtained for graded samples; removal of insects and foreign matter during cleaning ensures better and safer storage. In addition, pre-cleaned grain improves the efficiency of the drying machine during the drying process and results in a more uniformly dried sample. Some machines will simultaneously clean and grade a sample of grain or seed, while others will perform only one of these functions.

Powered

SEED CLEANERS/GRADERS TA RANGE
A range of five cleaners designed to suit the processing of grains and pulses into high quality seed.
Power source: electric
Alvan Blanch, UK

SEED CLEANERS/GRADERS TAE RANGE
For installations in which control of dust is a premium requirement. These machines are totally enclosed to minimize the emission of dust. The panels are either hinged or easily removable – allowing for full access for adjustment and maintenance. Perspex windows are included.
Power source: electric
Alvan Blanch, UK

SCREEN AIR SEPARATOR
Best suited as a pre-cleaner before the storage of seeds. Variable speed drive ensures accurate quality separations at high capacity. Totally enclosed aspiration system with cyclone provides dust-free working in the plant. Capacity: 100–2000 kg/hour
Power source: battery/electric
Price code: 3
Goldin (India) Equipment (Pvt) Ltd, India

SAB-PRE CLEANER AND UB FINE CLEANER
This machine is used for fine cleaning and grading.
John Fowler (India) Ltd, India

GRAIN CLEANING MACHINE
This machine is used for cleaning milled rice.
Capacity: 500 kg/hour
Power source: electric
Price code: 3
Kongsonglee Kanchang, Thailand

GRAIN CLEANER
Grain cleaning machine.
Marot, France

GRAIN–SEED–FLOUR CLEANER AND GRADER
Used for cleaning and grading various types of cereals. Capacity: 500 kg/hour
Power source: electric
Rajan Universal Exports (Manufacturers) Pvt Ltd, India

RICE POLISHING MACHINE RS 40
Moisture content of rice grains is increased by a humidifier before feeding between milling rollers for skinning and polishing. Rice bran is removed by centrifugal fans. Percentage of broken rice grains is less than 2 per cent; produces export-quality rice. Good design, easy and safe to operate. Automatic control on input and output. Weight: 1200 kg, dimensions: 2300 × 920 × 2800 mm. Capacity: 4 tonnes/hour
Power source: electric
Price code: 4
Sai Gon Industrial Corporation (SINCO), Vietnam

RICE WHITENER VSW 40
Removes bran by creating friction between rice grains and emery/grinding net (first stage) then between rice grains and blade/grinding net (second stage). Rice bran is removed by a centrifugal fan. Rice with high humidity (15–19 per cent) can be used. There is a low percentage of broken rice grains after two stages of whitening: 6–12 per cent. Weight: 1500 kg, dimensions: 1700 × 1400 × 2200 mm. Capacity: 3–4 tonnes/hour
Power source: electric
Price code: 4
Sai Gon Industrial Corporation (SINCO), Vietnam

Manual

HAND OPERATED GRAIN CLEANER
This machine is used for cleaning and grading of grains. It separates impurities such as stubble, chaff and dirt from the grain. Provided with three extra sieves to make it suitable for a variety of cereals and pulses.
Capacity: 150 kg/hour
Power source: manual
Central Institute of Agricultural Engineering, India

PEDAL-CUM-POWER OPERATED GRAIN CLEANER
This machine can be operated manually or by electric motor. It removes stubble, dirt and broken or small size grain to deliver clean graded grain. Capacity: 200–400 kg/hour
Power source: electric/manual
Central Institute of Agricultural Engineering, India

8.0 Cold storage equipment

Reducing the temperature is one way of extending the shelf life of perishable goods and raw materials. Refrigerators and chillers can be used during processing and also for storage of products such as soft drinks and juices, which are more refreshing when served cold. Refrigerators can be battery operated, gas powered, kerosene or run from mains electricity.

PORTABLE CHILLER
Ideal for chilling soft drinks and snacks.
Capacity: 40 litres
Power source: battery
Modern Erection, Bangladesh

8.1 Refrigerators

REFRIGERATION & AIR CONDITIONING EQUIPMENT
A full range of refrigeration and air conditioning equipment.
HRP Focus Ltd, UK

MINI CHILLER
This chiller can be used for chilling bottled and canned products as well as for storage purposes for fruits, bakery products and sweets etc. Capacity: 60 litres
Power source: electric
Modern Erection, Bangladesh

REFRIGERATOR
The refrigerator comprises two large original fruit containers with covers. Shelves on the door panel are adjustable. A closed shelf is provided for storing butter, cheese and other strong-smelling foodstuffs. The low temperature compartment can go as low as minus 12 degrees. Capacity: 155–160 litres
Power source: electric
Price code: 1
Odessa State Academy of Food Technology, Ukraine

8.2 Freezers

FT25A CONTINUOUS ICE CREAM FREEZER
Fully self-contained unit in mobile cabinet. The scraped surface unit is easily dismantled for cleaning and maintenance. Rapid cooling of viscous systems; ice cream freezing and air incorporation; iced desserts and yoghurts.
Capacity: 10 litres/hour
Power source: electric
Armfield Ltd, UK

FT36 BLAST AND FLUID BED FREEZER
A floor-standing, insulated freezer cabinet incorporating both blast and fluid-bed freezer capabilities. Access doors for both sections. Blast freezer section has five removable stainless steel trays.
Power source: electric
Armfield Ltd, UK

ICE CONFECTIONERY FREEZING MACHINE
Power source: electric
Price code: 1
Bay Hiap Thai Company Ltd, Thailand

Freezers and ice-making machines are also manufactured by Industrial Refrigeration Pvt Ltd, India.

9.0 Confectionery equipment

Specialized confectionery equipment includes depositors, enrobers, and moulding and coating machines. See also sections 24.0 (mixers) and 22.0 (ice cream equipment).

BAKERY AND CHOCOLATE MAKING EQUIPMENT
Depositors, enrobers and moulding machines available.
Charles Wait (Process Plant) Ltd, UK

LODIGE HI-COATER SYSTEMS
Suitable for film and sugar coating. Capacity: 0.75–550 litres
Gebruder Lodige Maschinenbau GmbH, Germany

10.0 Cutting, mincing and slicing equipment

A range of cutters, slicers and mincing machinery is available according to the type of raw material used and the scale of operation. Both manual and powered versions of the equipment are available.

MIXER, BLENDER AND SLICER
Performs the following functions: the mixing and blending of liquids; production of finger chips of potatoes and hard stoneless fruits; the slicing and shredding of potatoes, cabbage, carrots, apples etc.; the crushing of ice; kneading of flour, mince meat and fruits; the whisking and beating of eggs.
Power source: electric
Price code: 1
Gardners Corporation, India

Cutting, mincing and slicing equipment is also manufactured by Alven Foodpro Systems (P) Ltd, India, and Hariom Industries, India.

10.1 Fruit and Vegetable Cutters

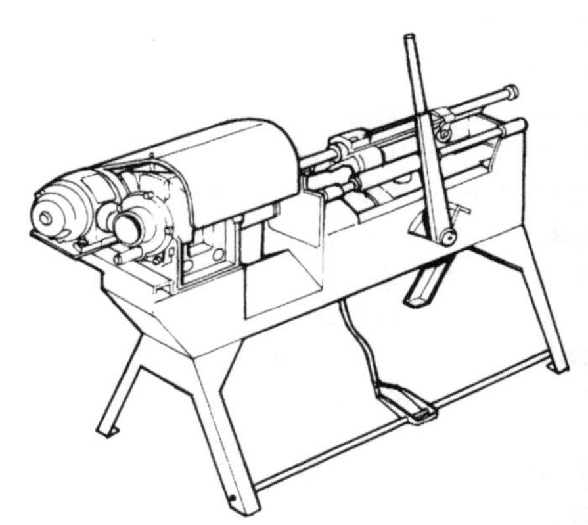

Powered

VEGETABLE SLICING MACHINE
The raw material drum lies horizontally with an input gutter. Cutting platform permits vertical slicing by three blades spaced at 120 degrees. Cutting blades can be adjusted to produce slices from 0.5 to 5 mm thickness. Continuous operation. The machine is made of stainless steel with easy to replace parts. Needs only one operator. Capacity: 20 kg/hour
Power source: electric
Price code: 3
ALFA Technology Transfer Centre, Vietnam

SLICING MACHINES
Used for the efficient slicing, cubing and shredding etc. of fruit.
Power source: electric
Alvan Blanch, UK

CHIPPER RCII
This machine cuts potatoes into chips, with a choice of ten sizes. Capacity: 1500 kg/hour
Power source: electric
Price code: 4
Crypto Peerless Ltd, UK

GARI PROCESSING MACHINE
This machine is used for processing *gari* (roasted cassava) on a small scale. Capacity: 50 kg/hour
Price code: 3
Department of Agriculture, Forestry and the Environment, Sierra Leone

VEGETABLE PREPARATION MACHINE
Specially designed for mounting onto a table. It grates, slices and cubes all types of vegetables. It has a 1 hp motor.
Power source: electric
Price code: 3
DISEG (Diseno Industrial y Servicios Generales), Peru

PINEAPPLE SLICER
Semi-automatic. Capacity: 100–300 kg/hour
Power source: electric
Eastend Engineering Company, India

POTATO/GINGER SLICER
Capacity: 25–100 kg/hour
Eastend Engineering Company, India

SLICING MACHINE
Adjustable thickness of slices. Shredder plate and finger cutter plate extra.
Power source: electric
Price code: 1
Gardners Corporation, India

BANANA CUTTER
1.2 hp/1700 rpm motor. Capacity: 250/300 kg/hour
Power source: electric
Price code: 3
Industrias Technologicas Dinamicas SA, Peru

FRUIT AND VEGETABLE SLICER
This slicer can be used for most fruits and vegetables. The thickness of slices can be adjusted.
Price code: 1
Narangs Corporation, India

VEGETABLE CUTTING MACHINE
Power source: electric
Rajan Universal Exports (Manufacturers) Pvt Ltd, India

VEGETABLE CUTTER
This equipment is suitable for home or group use.
Robot Coupe, France

VEGETABLE CUTTER
Capacity: 600 kg/hour
Power source: electric
Servifabri SA, Peru

VEGETABLE PREPARATION MACHINE
This machine has dies for cutting, slicing and grating vegetables. Capacity: 200 kg/hour
Power source: electric
Price code: 3
Servifabri SA, Peru

VEGETABLE DICER
Dicing machine suitable for a range of vegetables.
Power source: electric
Price code: 3
Servifabri SA, Peru

VEGETABLE PROCESSOR
Vegetable cutter made with an aluminium casing with stainless steel cutters. Can be used to chop, grate and dice a range of vegetables. Capacity: 200 kg/hour
Power source: electric
Price code: 3
Servifabri SA, Peru

Manual

POTATO SLICER
Manually operated equipment for making potato chips in small-scale processing. It accommodates potatoes up to 75 mm diameter. Capacity: 30–35 kg/hour
Power source: manual
Central Institute of Agricultural Engineering, India

PINEAPPLE CUTTER
Pedal-operated pineapple cutter cuts fruit into 10 mm rings.
Power source: manual
Price code: 2
DISEG (Diseno Industrial y Servicios Generales), Peru

FRUIT AND VEGETABLE SLICER
Hand-operated fruit and vegetable slicer. Arrangement to adjust the thickness of slices from paper thickness to 25 mm, to handle small as well as large fruits and vegetables.
Power source: manual
Price code: 1
Gardners Corporation, India

FRUIT AND VEGETABLE SLICING MACHINE
Operated with arrangement to adjust the thickness of slices from paper thickness to 2 cm.
Power source: manual
Price code: 1
Gardners Corporation, India

HAND OPERATED SLICER FOR FRUITS AND VEGETABLES
Light-duty hand-operated slicer for fruits and vegetables up to 7 cm in diameter.
Power source: manual
Price code: 1
Gardners Corporation, India

PINEAPPLE CUTTING KNIFE
Stainless steel cutting knife. Length 17 cm.
Power source: manual
Price code: 1
Gardners Corporation, India

PINEAPPLE PUNCH AND CORER
Pineapple punch to cut round slices to fit into cans, with a coring device.
Power source: manual
Price code: 1
Gardners Corporation, India

STAINLESS STEEL KNIVES
Fruit-cutting knives, coring knives, pitting knives and peeling knives.
Power source: manual
Price code: 1
Gardners Corporation, India

PINEAPPLE PUNCH
This hand-operated machine cuts pineapples into round slices for canning.
Power source: manual
Narangs Corporation, India

MANUAL POTATO CHIPPER
Power source: manual
Price code: 1
Servifabri SA, Peru

Fruit cutters are also manufactured by Bombay Industrial Engineers, India; Geeta Food Engineering, India; Hariom Industries, India; and Techno Equipments, India.

10.2 Meat cutters

FRESH MEAT SLICING MACHINE
Capacity: 100 kg/hour
Power source: electric
Price code: 2
Bay Hiap Thai Company Ltd, Thailand

GRAVITY FEED SLICERS
All models have a slice size adjuster and built-in knife sharpeners, and are belt driven with a manual feed carriage.
Crypto Peerless Ltd, UK

HANDAMATIC
Suitable for use as a pie cutting/moulding machine. Capacity: 400 pies/hour
Power source: manual
Price code: 4
Crypto Peerless Ltd, UK

BOWL CUTTER
A bowl cutter with stainless steel knives. Available in a range of sizes: 20 kg (with three knives), 40 kg and 90 kg (both with six knives). Available with two sizes of motor.
Capacity: 20 kg, 40 kg, 90 kg
Power source: electric
Servifabri SA, Peru

CUTTERS
Machinery for the sausage industry.
Capacity: 15–40 litres/hour
Talsabell SA, Spain

A meat cutter is also manufactured by Alven Foodpro Systems (P) Ltd, India.

10.3 Meat mincers

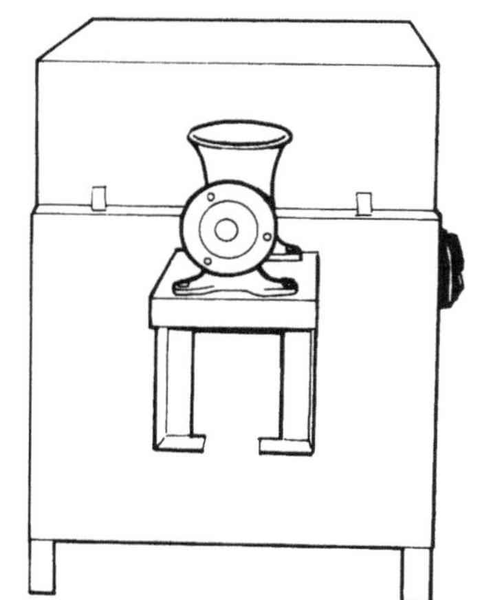

MINCERS
These machines mince fresh, tempered or frozen meat and have a continuous forward impulse/reverse switch. Capacity: 230–400 kg/hour
Power source: electric
Crypto Peerless Ltd, UK

MEAT MINCING MACHINE
Used to mince different kinds of meat into a paste. The capacity can be increased by fitting a higher powered motor. Capacity: 70 kg/hour
Power source: electric
Price code: 1
Dinh Chi Thang, Vietnam

MEAT MINCING MACHINE
Used to mince meat into a paste for sausage making etc. Capacity: 100 kg/hour
Power source: electric
Price code: 1
Doan Binh Mechanical Cooperative, Vietnam

MEAT MINCING MACHINE
Used to mince meat into a paste. Capacity can be varied to suit the customer by varying the size of motor. Capacity: 100 kg/hour
Power source: electric
Price code: 1
Duc Huan Mechanical Cooperative, Vietnam

GRINDING MACHINE
Capacity: 100 kg/hour
Power source: electric
Price code: 2
Eamseng, Thailand

MEAT MINCER
Capacity: 25–100 kg/hour
Eastend Engineering Company, India

MINI MEAT MINCER
Capacity: 75–125 kg/hour
Power source: electric/manual
Rajan Universal Exports (Manufacturers) Pvt Ltd, India

MEAT GRINDERS
A range of meat grinders and various attachments, such as stuffing tubes, plates, knives and other spare parts.
Power source: manual/electric
Price code: 1, 2, 3
Sausage Maker Inc., USA

MEAT GRINDERS
These meat grinders have single or three phase motors with a capacity of between 80 and 500 kg.
Power source: battery/electric
Price code: 3
Servifabri SA, Peru

MEAT GRINDER
These meat grinders are made of stainless steel and are used for grinding meat for the sausage industry. Available in a range of sizes with 1, 2 or 5 hp motor. All sizes available in single or tri-phase. Capacity: 80, 180 or 500 kg
Power source: electric
Price code: 3
Servifabri SA, Peru

GRINDERS/MINCERS
Machinery for the sausage industry.
Capacity: 700–1200 kg/hour
Talsabell SA, Spain

MEAT MINCER
Has a vertical shaft and is used for pork, beef, fish and cuttlefish. It also extrudes in round form for sausage making. Capacity: 100 kg/hour
Power source: electric
Price code: 2
Van Duc Cooperative, Vietnam

10.4 Pasta and noodle cutters

RICE PANCAKE CUTTER
Cuts rice pancakes, noodles, fine noodles or vermicelli made from cassava. The capacity can be expanded by fitting a motor. Capacity: 5–10 kg/hour
Power source: manual
Price code: 1
Bui Van Thanh Private Company, Vietnam

CUTTING MACHINE FOR PASTA
Used for cutting noodles, vermicelli from cassava and rice pancake. The capacity can be increased by fitting a motor.
Capacity: 10 kg/hour
Power source: manual
Price code: 1
Dinh Van Bay, Thailand

NOODLE SHEETING/CUTTING MACHINE
Capacity: 50 kg/hour
Power source: electric
Price code: 2
Thaworn Kanchang, Thailand

10.5 Cereal cutters

SLICERS
Power and gravity feed slicers to suit the requirements of small- and large-scale bakeries. Capacity: 200–800 loaves/hour
Power source: diesel/electric
Price code: 4
Baker Enterprises, India

BREAD SLICING MACHINE
Power source: electric
Bijoy Engineers, India

BREAD SLICING MACHINE
Capable of handling bread up to 25 cm long.
Power source: electric
Price code: 2
Gardners Corporation, India

POWER OPERATED DOUGH DIVIDER
Power source: electric
Price code: 2
Gardners Corporation, India

SM100 CUTTING MILL
The final fineness of the product depends on the bottom screen used (mesh sizes 0.25 to 20 mm). There are three knives inside the cutting chamber ensuring fast but gentle pulverization. Capacity: 5–12 kg/hour
Power source: electric
Glen Creston Ltd, UK

NORTHSON CHAFF CUTTER
Rigid frame offers stability, and large flywheel ensures easy operation. Length of chop can be easily altered. Capacity: up to 400 kg/hour
Power source: manual/electric
G. North & Son (Pvt) Ltd, Zimbabwe

Bread slicers are also manufactured by Alven Foodpro Systems (P) Ltd, India, and Technoheat Ovens and Furnaces Pvt Ltd, India.

10.6 Oil-seed and nut cutters

COPRA CUTTER
Used for cutting coconuts.
Azad Engineering Company, India

KOMET CUTTING MACHINE CRUSHER
Used to reduce the size of nuts and other raw materials so that they can be more easily processed. The system prevents early oil-yield loss. Capacity: 100–300 kg/hour
Power source: electric
IBG Monforts GmbH & Co., Germany

COCONUT JELLY DICING MACHINE
Capacity: 100 kg/hour
Power source: electric
Price code: 2
Kasetsart University, Thailand

COPRA CUTTER
Cuts and breaks coconut balls into small pieces of 12 mm or less. Capacity: 150 kg/hour
Power source: electric
Price code: 2
Tinytech Plants, India

11.0 Dairy equipment

Dairy equipment includes all utensils that are required for the collection of milk through to the processing of butter, cream, cheese and yoghurt. Both hand-operated and powered machines are available, catering for all scales of operation.

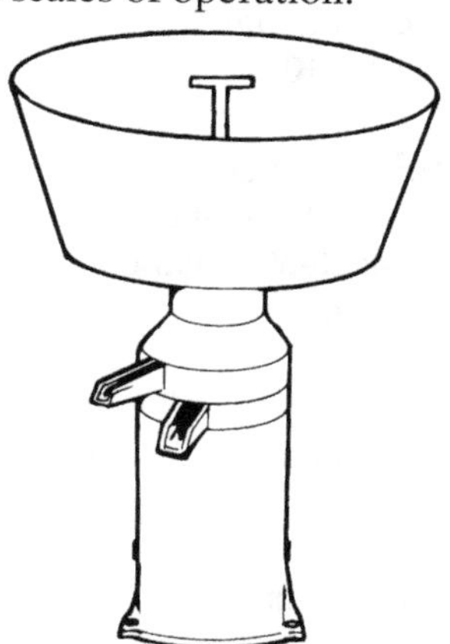

MILK BOTTLES AND BOTTLING EQUIPMENT
Batch sterilizer having a capacity of 50 bottles per batch and fitted with pressure gauge, safety valve, release valve and temperature gauge.
Power source: electric
Price code: 3
Dairy Udyog, India

MILK TESTING EQUIPMENT & UTENSILS
Includes hand-operated and electric centrifuge machines, funnels, beakers, flasks and milk collecting trays, cans and pails, as well as testing equipment such as butyrometers, hydrometers and filter cloths.
Dairy Udyog, India

DAIRY BY-PRODUCT MACHINERY AND CHILLERS
Includes freezers, boilers, chilling units, coolers, ghee and cheese vats, bottle sterilizers, cheese strainers and knives.
Dairy Udyog, India

MILK PROCESSING EQUIPMENT
A range of equipment for milk processing and on-farm milk products.
Elecrem, France

CONTAINERS/CANS/BUCKETS/PAIL
Capacity: 25–50 litres
Power source: manual
Fullwood Ltd, UK

CREAM SEPARATORS
Capacity: 80–500 litres/hour
Power source: electric/manual
Fullwood Ltd, UK

HANDI MILKER
Suitable for use as a mobile unit or a fixed unit linked to a vacuum line. Capacity: 18 litre tank
Power source: electric
Fullwood Ltd, UK

MINI MOBILE
Easy to manoeuvre unit. The carriage is designed to hold one milk can or bucket.
Capacity: 8 litre tank
Power source: electric/petrol
Fullwood Ltd, UK

SMALL-SCALE DAIRY TECHNOLOGY
A range of dairy farming equipment including fodder production, animal feeding, milk sanitation and milk processing equipment.
Small Scale Dairy Technology Group, The Netherlands

DAIRY CENTRIFUGE
The simplest type of equipment is the tubular bowl centrifuge that comprises a vertical cylinder or bowl, which rotates inside a stationary casing at between 1500 and 1600 rpm. The milk is introduced continuously at the base of the bowl and the two liquids (cream and skimmed milk) are separated into layers which emerge from the two outlets. A variety of disc-bowl centrifuges are also available. In these, a stack of conical discs inside a rotating bowl is used to separate the cream.

Dairy centrifuges are available from Dairy Udyog, India; Elecrem, France; Charles Wait Ltd, UK; and Lehman Hardware and Appliances, USA.

Dairy equipment is also manufactured by Elimeca SA, France, and Geere SA, France.

11.1 Butter making equipment

BUTTER CHURN – TYPE URC
Capacity: 150–600 litres total volume
Power source: electric
APV Unit Systems, Denmark

BUTTERMILK TROLLEY – TYPE VRH
Used for collection of buttermilk from conventional churns and for circulation of cleaning agents during the cleaning of the churn. Capacity: 50 litres
APV Unit Systems, Denmark

BUTTER-MAKING EQUIPMENT
A range of butter churns and butter making equipment for small- and large-scale producers.
Charles Wait (Process Plant) Ltd, UK

PASTEURISER–CHURN–KNEADER
Hot and cold water can be circulated between the two inner walls of this triple walled butter churn, enabling the milk to be both heated and cooled. The churn is provided with a variable speed motor for

churning the milk and kneading the butter.
Capacity: 50–400 litres
Power source: electric
C. van 't Riet Zuiveltechnologie BV, The Netherlands

BUTTER-MAKING MACHINERY
Includes butter churns, moulds, vats and balancing equipment.
Dairy Udyog, India

BUTTER CHURNS
Capacity: 7–32 litres
Power source: electric
Fullwood Ltd, UK

CREAM SEPARATOR AND BUTTER CHURN
The separator is equipped with a bearing and flywheel for easy hand turning. A hidden gear drive builds up tremendous centrifugal force to spin away cream. The internal flywheel keeps turning for over 30 seconds without being cranked, so one person can both crank and pour in milk. Capacity: 50 litres/hour
Power source: manual
Price code: 2
Lehman Hardware & Appliances, USA

Butter churns are also manufactured by Goma Engineering Pvt Ltd, India.

11.2 Cheese making equipment

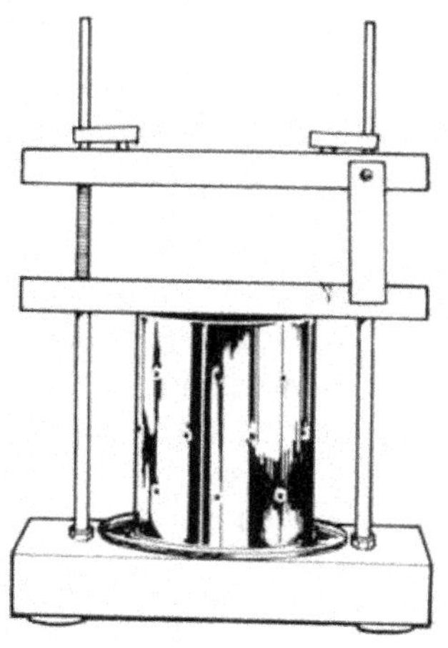

AGROLACTOR
Compact and automated platform for soya milk production. Capacity: 250 litres/hour
Power source: electric
Actini, France

FT20A CHEESE-MAKING ACCESSORIES
For use in conjunction with 10 litre cheese vat to produce approximately 1 kg of cheese. Items include: horizontal and vertical blade curd knives; Cheddaring box; cheese press; and cheese moulds.
Armfield Ltd, UK

FT20 CHEESE VAT
This jacketed rectangular vat can be used for the manufacture of all well-known cheese types. It has a removable paddle agitator of stainless steel, driven through a guarded linkage system by a variable speed drive motor. Capacity: 10 litres total volume
Power source: electric
Armfield Ltd, UK

COTTAGE SCALE SOYPANEER PLANT
A cottage-level soya paneer plant with machines including steam generation unit, grinder-cum-cooker, milk filtration unit and paneer pressing device. Capacity: 50 kg/day
Power source: manual
Central Institute of Agricultural Engineering, India

CHEESE-MAKING EQUIPMENT
A range of cheese-processing equipment including cheese vats, mould fillers, cutters and curd mills.
Charles Wait (Process Plant) Ltd, UK

HOT WATER SUPPLY UNIT
Complete hot water unit consisting of a gas- or oil-fired central heating boiler and a separate boiler for supplying hot running tap water. Designed for circulating hot water through cheese-making equipment, pasteurizer or churn.
C. van 't Riet Zuiveltechnologie BV, The Netherlands

PLASTIC CHEESE MOULD

An easy to handle cheese mould consisting of tub, tub net, cover and cover net. Capacity: 0.2–15 kg

C. van 't Riet Zuiveltechnologie BV, The Netherlands

PNEUMATIC CHEESE PRESS

Available with three or more cylinders with double pressure plates. The pressure in each cylinder can be separately adjusted and the cylinders can be operated independently of each other.

Power source: electric

C. van 't Riet Zuiveltechnologie BV, The Netherlands

STAINLESS STEEL CHEESE VAT

Hot and cold water can be circulated between the two inner walls of this circular cheese vat, enabling the milk to be both heated and cooled. Glass fibre insulation is fitted between the two outer walls ensuring both minimum energy use and maximum temperature stability. In addition the vat can be fitted with electrical heating elements which can raise the temperature high enough for pasteurization. Capacity: 200–1750 litres

Power source: electric

C. van 't Riet Zuiveltechnologie BV, The Netherlands

CHEESE PRESS

This is a portable, easily managed cheese press with stainless steel moulds. Four metal moulds can be used at one time. The moulds have holes so that water can be expelled easily by exerting pressure regulated by a vertical screw rod.

Power source: manual

Price code: 1

Intermediate Technology Development Group, Bangladesh

CHEESE PRESSES

Cheese presses remove the liquid from milk curd, which is essential for all hard cheeses. Wrapped curd is placed in the mould, covered with a pressing disc and left to drain.

Power source: manual

Price code: 1

Lehman Hardware & Appliances, USA

Cheese-making equipment is also manufactured by Brasholanda SA – Equipamentos Industriais, Brazil; Elimeca SA, France; and Geere SA, France.

11.3 Yoghurt-making equipment

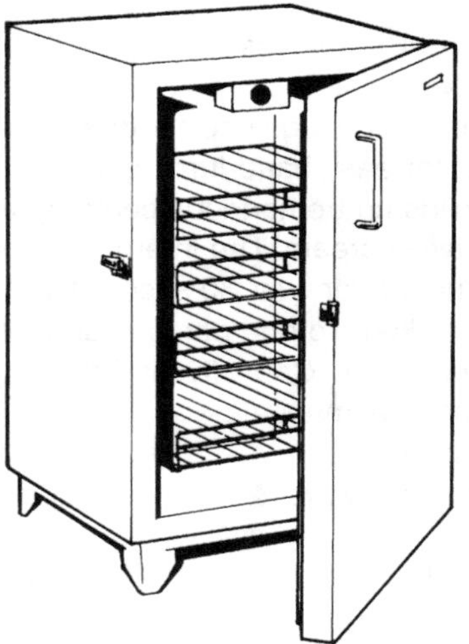

YOGHURT INCUBATOR

Capacity: 1000 cups/day

Power source: electric

Price code: 2

Ashoka Industries, Sri Lanka.

YOGHURT INCUBATOR

This incubator can be used to make yoghurt from any milk including cow, soybean or goat milk. It is light, portable and easy to clean. A super-insulated styrofoam canister retains warmth evenly for hours. The incubator can also be used for cheese making.

Power source: manual

Price code: 1

Lehman Hardware & Appliances, USA

12.0 Decorticators and de-hullers

Decorticating or de-hulling involves the removal of the outer shell or seed coat from cereals, seeds, nuts, pulses and oil-seeds. Some machines incorporate a blower, which separates the skin from the kernel. In others, decortication must be followed by manual winnowing to remove the bran and husk.

A range of machinery is available with various throughputs and capacities, and in both manual and motorized versions.

Powered

PALM NUT CRACKER
Used to crack the nut in order to obtain the kernel. Capacity: 750 kg/hour
Power source: diesel/electric
Agricultural Engineers Ltd, Ghana

MULTI CROP DECORTICATORS
Suitable for the removal of bran and germ from maize. Also suitable for sorghum, millet and rices. The size of screen is interchangeable to suit the type of grain to be processed. Capacity: 375–750 kg/hour
Power source: electric
Price code: 4
Alvan Blanch, UK

HULLER POLISHERS
Used for the complete milling of paddy into polished white rice. The cylindrical huller comprises a specially chilled and hardened cast iron cylinder which rotates inside a screen to produce shelled and whitened paddy in one operation. The degree of milling is controlled by the adjustable huller blade. A fan mounted on the side of the polisher blows the husk/bran away from the machine.
Capacity: 350–700 kg/hour
Power source: electric
Price code: 3
Alvan Blanch, UK

PALM NUT CRACKER: STANDARD
Used to efficiently crack the shells of palm kernel nuts. A centrifugal thrower rotates within a heavy gauge steel case. The machine is fitted with an input hopper and discharge chute. It is mounted on a single axle with two pneumatic tyred wheels and a hand tow bar. Capacity: 500 kg/hour
Power source: electric
Price code: 4
Alvan Blanch, UK

MOTORISED PULSES DEHUSKING MACHINES
AMI Engineering, India

DECORTICATOR
Used for the decortication of groundnuts.
Capacity: 10 kg/hour
Azad Engineering Company, India

HIGH CAPACITY GROUNDNUT DECORTICATOR
Power-operated equipment for shelling groundnut pods for edible and seed purpose. Breakage limited up to 2 per cent. Suitable for medium-scale commercial units. Capacity: 300–350 kg/hour
Power source: electric
Central Institute of Agricultural Engineering, India

SEED DECORTICATOR MACHINE
Capacity: 500 kg/hour
Power source: electric
Price code: 3
Flower Food Company, Thailand

DECAR 300 SHELLING MACHINE
Designed to shell groundnuts on an industrial scale. Also sorts out the unshelled elements and removes shells and dust by vacuum extraction. Capacity: 150–2000 kg/hour
Gauthier, France

IMPACT HULLER
Control of speed and feed permits operator to obtain best possible hulling of given material. Cyclone collector and closed air circulation provide dust-free working in the plant. Capacity: 100–2000 kg/hour
Power source: battery/electric
Price code: 3
Goldin (India) Equipment (Pvt) Ltd, India

MAIZE HULLERS
Capacity: 500 kg/hour
Power source: electric/diesel
H.C. Bell & Son Engineers (Pvt) Ltd, Zimbabwe

BARLEY DEHULLER
24 hp motor. Capacity: 450–500 kg/hour
Power source: electric
Price code: 4
Industrias Technologicas Dinamicas SA, Peru

SHELLING MACHINE
Power source: diesel/gas
Price code: 2
Lim Chieng Seng Ltd, Thailand

GROUNDNUT (PEANUT) DECORTICATOR
For de-shelling peanuts, using the blower to remove the broken shell. Capacity: 100–750 kg/hour
Power source: electric
Rajan Universal Exports (Manufacturers) Pvt Ltd, India

MAIZE SHELLER
Used to separate the maize kernels from the maize cobs. Capacity: 400–2000 kg/hour
Power source: electric
Rajan Universal Exports (Manufacturers) Pvt Ltd, India

MULTI CROP HULLER
Used to remove the outer skin on cereal kernels and polish them. Rice and maize can also be de-hulled in the same machine.
Capacity: 60–500 kg/hour
Power source: diesel/electric
Rajan Universal Exports (Manufacturers) Pvt Ltd, India

CASHEW SHELLING MACHINE
Motor-driven shelling machine. Capacity: 1200 nuts/hour
Power source: electric
Price code: 3
Sri Lanka Cashew Corporation, Sri Lanka

GROUNDNUT DECORTICATOR
This machine shells the groundnut and a built-in blower throws away husks while a shaking screen separates out good whole kernels and split kernels. Capacity: 300 kg/hour
Power source: electric
Price code: 2
Tinytech Plants, India

SUNFLOWER/PALM NUT CRACKER
This machine shells seeds by centrifugal force. Capacity: 300 kg/hour
Power source: electric
Price code: 2
Tinytech Plants, India

PEANUT SHELLER
This machine shells and separates the nuts, using a 3–4 kW electric motor. Capacity: 250 kg/hour
Power source: electric
Udaya Industries, Sri Lanka

BARLEY HULLER
A powered machine for de-hulling barley with a three phase 9 hp motor. Capacity: 1000–2000 kg/hour
Power source: electric
Vulcano Technologia Aplicada Eirl, Peru

GROUNDNUT SHELLERS
Used for removing groundnuts from outer shells efficiently and with minimum nut breakage. Capacity: 50–12 000 kg/hour
Power source: electric/manual
Price code: 2/3
Alvan Blanch, UK

MAIZE SHELLERS
Used for separating grains from cobs. Larger models also de-husk and handle sorghum/millet. Capacity: 50–12 000 kg/hour
Power source: manual/electric
Price code: 1/2/3
Alvan Blanch, UK

SOYBEAN DEHULLER
A unit specially designed to meet soybean de-husking and splitting requirement. Suitable for cottage to small-scale soybean processing units engaged in producing soya flour, milk, paneer and other soya products. Capacity: 100 kg/hour
Power source: electric/manual
Central Institute of Agricultural Engineering, India

Manual

HAND HULLER
This small machine is intended for use by small growers, milling rice for their own needs. The end product requires hand winnowing to remove bran and husk from the sample. Capacity: up to 15 kg/hour
Power source: manual
Price code: 2
Alvan Blanch, UK

HAND OPERATED PULSES DEHUSKING MACHINES
Power source: manual
AMI Engineering, India

GROUNDNUT DECORTICATOR
Suitable for small farmers to shell groundnut pods for seed and other purposes. Easily relocated. By using sieves of a smaller slot size it can be used for shelling castor as well. Capacity: 50–60 kg/hour
Power source: manual
Central Institute of Agricultural Engineering, India

SUNFLOWER DECORTICATOR
Used for the removal of husk of sunflower seed. Capacity: 10 kg/hour
Power source: manual
Central Institute of Agricultural Engineering, India

TUBULAR MAIZE SHELLER
A useful lightweight and simple hand tool for maize-growing small farmers. Breakage of grains or cobs is nil. Highly useful for shelling seed maize as viability of grain is not affected. Capacity: 15–25 kg/hour
Power source: manual
Central Institute of Agricultural Engineering, India

CORN SHELLER
This machine has self-contained clamps with a spring adjustment to fit all sizes of corn. Cob ejector and tipping attachment are included.
Power source: manual
Price code: 1
C.S. Bell Co., USA

'CROCODILE' HAND DRIVEN MEALIE SHELLER
Designed for home use.
Capacity: 270 kg/hour
Power source: manual
G. North & Son (Pvt) Ltd, Zimbabwe

HUNTSMAN 2 HOLE SHELLER
This machine grades into two sizes of maize and is adjustable for the cob size. Power pulley is also available. Capacity: 10 bags per hour
Power source: manual
G. North & Son (Pvt) Ltd, Zimbabwe

CASHEW CRACKER
This is a simple, low-cost cracker with blades that split the outer shell without breaking the nut inside.
Power source: manual
Kaddai Engineering, Ghana

KERNEL DESHELLING MACHINE
Capacity: 50 kg/hour
Power source: manual
Price code: 1
Kasetsart University, Thailand

CASHEW NUT DECORTICATOR
Before being decorticated, the cashew nuts must be dried in the sun for at least three days and then placed into boiled water for five minutes before being dried in the sun for a further one or two days.
Capacity: 5–10 kg/man/day
Power source: manual
Kunasin Machinery, Thailand

PADDY TYPE GROUNDNUT STRIPPER
Capacity: 10–15 kg/man/hour
Power source: manual
Kunasin Machinery, Thailand

RUBBER TYRE GROUNDNUT SHELLER
Capacity: 40–60 kg/hour
Power source: manual
Kunasin Machinery, Thailand

CASHEW DECORTICATOR
Capacity: 16 kg/hour
Power source: manual
Rajan Universal Exports (Manufacturers) Pvt Ltd, India

MAIZE SHELLER – SHELL MASTER TWIN FEED
This machine separates the maize kernels from its cob. It can be converted to power operation. Capacity: 150–200 kg/hour
Power source: manual
Rajan Universal Exports Pvt Ltd, India

TWO MAN HAND HULLER
Capacity: 40–250 kg/hour
Power source: manual
Rajan Universal Exports (Manufacturers) Pvt Ltd, India

DECORTICATOR
This machine removes the skin from groundnuts etc. Capacity: 50 kg/hour
Power source: manual
Udaya Industries, Sri Lanka

Hullers are also manufactured by B Sen Barry & Co., India, and Electra, France.

12.1 Rice hullers

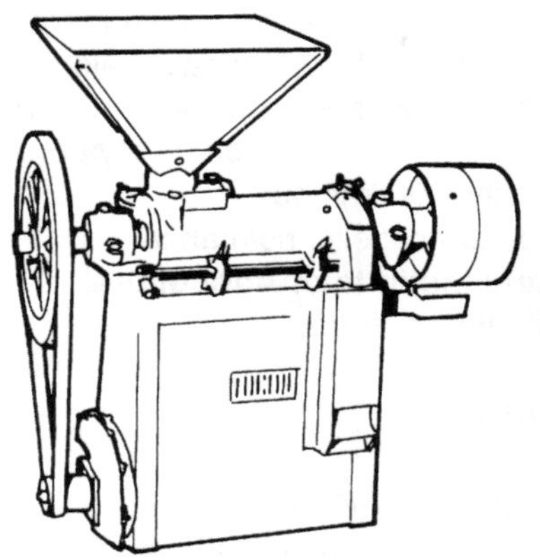

Powered

PADDY DEHUSKERS
Used for de-husking rice and seeds.
Capacity: 250–2500 kg
Power source: electric/diesel
DIW Precision Engineering Works, India

DBR 800 HUSKING AND WHITENER MACHINE
This machine is used for husking paddy and whitening brown rice.
Gauthier, France

PADDY HULLER
This machine is used for hulling, separating and polishing rice. It is available in a range of models with between 4 and 12 kW electric motors. Capacity: 250–2000 kg/hour
Power source: electric
Udaya Industries, Sri Lanka

RICE HULLER
Capacity: 100–500 kg/hour
Price code: 4
Votex Tropical, The Netherlands

PADDY HULLING MACHINES
Capacity: 1000 kg/hour
WA ITTU, Ghana

12.2 Coffee hullers

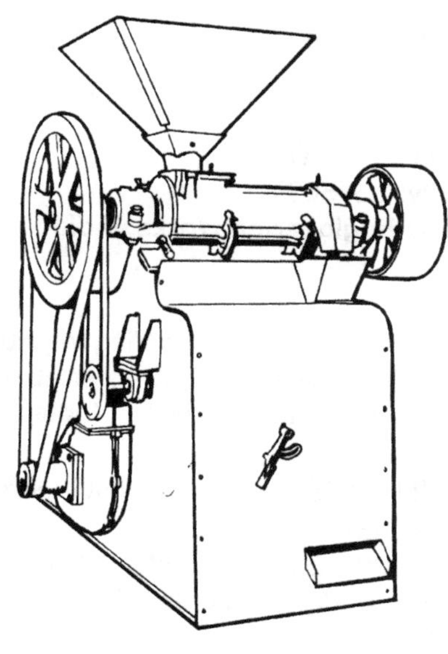

COFFEE HULLER
Used for the hulling of coffee. All hullers are fitted with hulling knives and screen clamps in stainless steel.
Power source: diesel/petrol/electric
Price code: 4
Alvan Blanch, UK

HUSKING MACHINE – DQC 500
Suitable for hulling robusta coffee (dry method) or for parchment removing from arabica coffee (wet method). Other seeds can also be treated, e.g. groundnuts, almond.
Capacity: 500–1200 kg/hour
Gauthier, France

LIMPRIMITA COFFEE HULLER
Any quantity from one bean to several kilograms can be shelled rapidly, completely and safely using this machine. A ribbed roller operates in conjunction with an adjustable flexible block.
Power source: electric/manual
Price code: 2/3
John Gordon International, UK

Manual

BUKOBA COFFEE HULLER
Shells dry cherry or parchment coffee. The coffee is simply fed to the machine and the handle turned. The shelled coffee is discharged at the end of the machine mixed with the shells and dust, and must then be winnowed and sieved. Capacity:
9–27 kg/hour
Power source: manual
Price code: 2
John Gordon International, UK

13.0 De-stoners

De-stoners are used to remove the stone or seed from the centre of fruits prior to processing.

CHERRY PITTING MACHINE
Comes with a feed hopper.
Power source: manual
Price code: 1
Gardners Corporation, India

14.0 Drying equipment

Drying foods in the sun is probably the oldest method of food preservation. At the simplest level, the food is spread on the ground or on mats and the moisture at or near the surface of the food is heated and vaporized by the heat from the sun and ambient air. However, there are a number of disadvantages with sun-drying. These include: the intermittent nature of solar energy throughout the day and different times of the year; the possible contamination of the food material by dirt or insects; exposure to the elements (such as rain and wind), causing spoilage and losses; and the exposure of the crops to rats, chickens and other pests.

Most solar dryers incorporate a platform raised above the ground and often some kind of covering to combat the problems mentioned above. The purchase of a solar dryer may require a high capital investment initially, but it is argued that the gains from producing a higher quality product will quickly offset these costs. A careful assessment of the requirements of the producer is therefore necessary to establish whether higher income can be obtained from better quality products. Solar dryers may be more suitable if processing high-value foods rather than low-value staple crops. In addition, the introduction of combination solar/fuel-fired dryers may be a desirable option to overcome the problems of intermittent sunshine that often adversely affect solar drying.

Vacuum dryers and drum dryers are more costly and are generally used for larger-scale processing operations.

Tray dryers have proved successful in drying many products, particularly the higher value foods such as fruit and herbs. The capacity of the dryer varies according to the product being dried, but as an example it will dry 300–400 kg of herbs (net weight) per day. The heat source can be provided by diesel, gas, electricity or biomass. The dryer requires 1200 cubic feet of hot air per minute.

Detailed construction plans and details of the ITDG tray dryer are available from ITDG.

FLUID BED DRYER
Capacity: 2.5–500 kg
Bombay Engineering Works, India

DRYERS
A wide range of dryers, including tray, tunnel, drum and fluid bed dryers, are available.
Bombay Industrial Engineers, India

CONTINUOUS GRANULATING DRYER DRUVATHERM
Capacity: 300–2000 litres total volume
Gebruder Lodige Maschinenbau GmbH, Germany

GAS DRYING MACHINE
Capacity: 30 kg/hour
Power source: gas
Price code: 2
Narongkanchang, Thailand

HOT AIR DRYER
For uniform circulation of hot air in the drying chamber.
Rank and Company, India

Dryers are also manufactured by Sri Murugan Industries, India.

14.1 Solar dryers

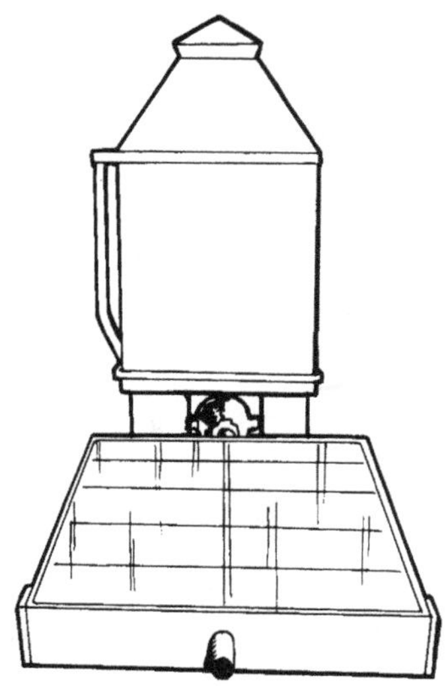

SOLAR DRYER
Power source: solar
AGCM (Atelier General de Construction Metallique), France

SOLAR DRYER
Capacity: 15 kg in two days
Power source: solar
Price code: 2
ASELEC (Atelier de Soudure et d'Electricité), Burkina Faso

SOLAR DRYER
Power source: solar
Atelier KONATE B Boubacar, France

SOLAR DRYER
Power source: solar
CEAS-ATESTA (SAPE. SATA), Burkina Faso

SOLAR CABINET DRYER
Uses solar energy for the drying of perishables, semi-perishables and wet processed food materials. No mechanical or electrical power is required and it is free from the hazard of fire. A specially designed aspirator improves air flow to reduce drying time. Capacity: 30–50 kg/batch = 5–7 days' drying time.
Power source: solar
Central Institute of Agricultural Engineering, India

SOLAR TUNNEL DRYERS
A range of dryers, suitable for many commodities and designed for small professional enterprises.
Innotech, Germany

SOLAR DRYER
Power source: solar
Kabore Koutiga Jean et Frères (Etablissements), Burkina Faso

FOOD SOLAR DRYING MACHINE
Capacity: 50 kg/24 hours
Power source: solar
Price code: 2
Kasetsart University, Thailand

SOLAR DRYER
Power source: solar
Kinate et Frères (Etablissement), Burkina Faso

CABINET SOLAR DRYER
Power source: solar
Natural Resources Institute, UK

SOLAR DRYER
Power source: solar
Soldev – Soleil et Developpement, Burkina Faso

TENT SOLAR DRYER
Power source: solar
Price code: 1
Technology Consultancy Centre, Ghana

14.2 Fuel-fired dryers

DRYER
Uses paddy husk or sawdust as fuel.
Capacity: 15–75 kg/10 hours
Power source: biomass
Price code: 2
Ashoka Industries, Sri Lanka

FOOD DRYER
Made of a wooden cabinet, with trays made of plastic mesh. Air is extracted by a fan, which has a 2.4 hp motor (1750 rpm).
Capacity: 300 kg herbs dried in 10 hours
Power source: diesel/electric
Industrias Technologicas Dinamicas SA, Peru

DIESEL/KEROSENE DRYER
Has 60 (4' × 2') trays.
Power source: diesel
Price code: 4
Kundasala Engineers, Sri Lanka

14.3 Electric dryers

ELECTRIC DRYER
Capacity: 10–40 kg/10 hours
Power source: electric
Price code: 2
Ashoka Industries, Sri Lanka

MULTIPURPOSE DRYER FOR SOY BEAN PRODUCTS
Moisture removal of between 40 and 50 per cent is achievable. Capacity: 250 kg/5 hours
Power source: electric
Central Institute of Agricultural Engineering, India

PADDY DRYERS
Used for drying seed. Capacity: 250–2500 kg
Power source: electric/diesel
DIW Precision Engineering Works, India

FOOD DRYING MACHINE
Capacity: 50 kg/hour
Power source: electric
Price code: 2
Kasetsart University, Thailand

ELECTRIC DRYER
This dryer has 15 (4' × 2') trays.
Power source: electric
Price code: 3
Kundasala Engineers, Sri Lanka

RECIRCULATING GRAIN DRYER
Equipment can be adjusted to buyer's needs.
Capacity: 1000–5000 kg/hour
Power source: electric
Price code: 4
Research Institute for Agricultural Machinery, Vietnam

DRYERS
Multipurpose dryers available for a wide range of food and agricultural products.
Capacity: 100–400 kg/10 hours
Power source: electric
Udaya Industries, Sri Lanka

14.4 Vacuum dryers

DOUBLE CONE DRYER
The material to be dried is tumbled in a slow rotating vessel of double cone construction which is jacketed for steam, hot water or oil heating. As the dryer rotates, the rapid and thorough intermixing of the entire batch brings every particle in contact with the heated metal. These dryers are suitable for damp powders or granules. Capacity: 28–10 500 litres
Mitchell Dryers Ltd, UK

VACUUM SHELF DRYER
Wet material is carried on trays which are placed onto heating plates within the vacuum chamber. High-efficiency heating plates in direct contact with the total tray surface ensure maximum heat transfer and even drying. Circular or rectangular units can be supplied. Capacity: sizes 1.25–19 square metres
Mitchell Dryers Ltd, UK

14.5 Drum dryers

BATCH DRYER
An efficient and economical method for processing a wide variety of crops. Used for the drying of grains, pulses and non-granular materials such as alfalfa, herbs and root crops – cassava, yams, chillies etc. Capacity: 1500–7000 kg/hour
Power source: electric
Price code: 4
Alvan Blanch, UK

DRUM DRYING MACHINE
Capacity: 1 kg/hour
Power source: electric
Price code: 3
Kasem Kanchang, Thailand

ROASTING–DRYING MACHINE MSR-500
This is a rolling-drum drying machine. There are helical blades inside the drum which lies horizontally with bearings at both ends.
Capacity: 300 kg/hour
Power source: electric
Price code: 3
Mechanical Engineering Faculty, University of Agriculture, Vietnam

FILM DRUM DRYERS/FLAKERS
Provide a means of continuously drying suspensions and solutions on a small to medium scale. Moisture is evaporated and the product scraped off as a powder or flake. These dryers are available in single or double drum configurations. Single drum dryers can also be used for chilling and flaking molten materials utilizing a coolant applied to the interior of the drum. The molten feed is applied using a heated trough or roller. After solidification it is scraped off, usually as a flake. Capacity: up to 1000 kg/hour
Power source: electric
Mitchell Dryers Ltd, UK

14.6 Tray dryers

TRAY DRYER
Used for removing moisture from food products such as spices etc. Contains 12 trays. Capacity: 50 kg/charge
Power source: electric
Price code: 3
Acufil Machines, India

TRAY DRYER
A range of dryers are available, containing from 6 to 384 trays. Special custom-built dryers, conveyor dryers etc. are also available.
Bombay Engineering Works, India

CABINET BATCH DRYER
Made of wood, measuring 2.4 m (height) by 1.2 m (depth) by 3.6 m (length). It is fitted with 65 wooden trays with nylon mesh. Each tray measures approximately 70 × 115 cm. It is powered by an automatic kerosene heater with heat exchanger. Capacity: 300 kg/batch
Price code: 3
Industrias Technologicas Dinamicas SA, Peru

TRAY DRYERS
Trays of wet material are loaded directly into the cabinet dryers. Capacity: 3.3–240 square metres
Power source: electric
Mitchell Dryers Ltd, UK

Tray dryers are also manufactured by Premium Engineers Pvt Ltd, India; Shirsat Electronics, India; and Sri Venkateswara Industries, India.

15.0 Expellers

Expellers, which are similar to extruders, are used for extracting oil from oil-seeds or nuts. They consist of a horizontal cylinder, which contains a rotating screw. The seeds or nuts are carried along by the screw and the pressure is gradually increased. The material is heated by friction and/or electric heaters.

The oil escapes from the cylinder through small holes or slots and the press cake emerges from the end of the cylinder. Both the pressure and temperature can be adjusted for different kinds of raw material.

Powered

'JAGDISH' TINY OIL PRESS
Used for the extraction of vegetable oil from oil-seeds. The reverse worm system in the pressing chamber results in the highest recovery of oil. Capacity: 50 kg/hour
Power source: diesel/electric
Price code: 3
Agro Industrial Agency, India

OIL SCREW PRESS – MODEL MINI 50
Barrel constructed from separate cast rings spaced apart by shims. The barrels are 60 mm diameter by 155 mm drainage length, with a single piece worm shaft driven from the discharge end through a totally enclosed intermediate chain. The unit is complete with feed hopper, manual feed chute, oil discharge spout and base plate. A versatile, low-cost machine, designed to suit village communities or small industries. Suitable for oil and cattle cake. Capacity: 50 kg/hour input
Power source: diesel/electric
Price code: 4
Alvan Blanch, UK

MM2 SCREW PRESS
This machine can be used to separate creamed desiccated coconut into virgin coconut oil and a fine aromatic desiccated coconut. It is also used for other fruits and vegetables.
Power source: electric
Arai Machinery Corporation, Japan

OIL EXPELLER
Capacity: 5 litres/hour
Power source: electric
Price code: 2
Ashoka Industries, Sri Lanka

'AZAD' BRAND OIL EXPELLER
Used for the extraction of oil from all kinds of oil-bearing seeds. Capacity: 10 kg/hour
Power source: electric
Azad Engineering Company, India

KOMET SINGLE SCREW VEGETABLE OIL EXPELLER
For the extraction of vegetable oil by cold pressing from oil-seeds, kernels and nuts.
Capacity: 13–35 kg/hour
Power source: electric
IBG Monforts GmbH & Co., Germany

EXPELLER
This machine is used for crushing palm kernels, groundnut seeds and moringa seeds to extract oil. Capacity: 5000 litres/hour
Power source: electric/manual
Price code: 3
Nigerian Oil Mills Ltd, Nigeria

OIL EXPELLER
Capacity: 15–210 kg/day
Power source: electric
Rajan Universal Exports (Manufacturers) Pvt Ltd, India

PEANUT BUTTER MAKING MACHINE
This machine is used for making butter from peanut, almonds, walnuts etc. Capacity: 50–200 kg/hour
Power source: electric
Rajan Universal Exports (Manufacturers) Pvt Ltd, India

TEMDO OIL EXPELLER EXP. 30
This machine is suitable for the production of oil and animal feed cake from oil-seeds.
Capacity: 180–200 kg/hour of seeds
Power source: diesel/electric
TEMDO (Tanzania Engineering and Manufacturing Design Organisation), Tanzania

OIL EXPELLER
For extracting oil from any oil-seed. Capacity: 125 kg/hour
Power source: electric
Price code: 3
Tinytech Plants, India

OIL GHANI
Suitable for the extraction of oil from oil-seeds using traditional pestle and mortar method. Capacity: 12 kg per batch
Power source: electric
Price code: 2
Tinytech Plants, India

HAND OIL PRESS
Used for the extraction of oil from a wide range of oil-bearing crops. Robust construction with 17 litre capacity perforated barrel. Two levers (for two person operation). Collection tray for pressed seed and outlet for oil. Capacity: 100 litres of oil after 6 hours of pressing.
Power source: electric/manual
Price code: 4
Alvan Blanch, UK

Manual

KOMET SINGLE SCREW OIL EXPELLER
Hand-operated machine with base plate.
Capacity: 2–8 kg/hour (seed to be extracted)
IBG Monforts GmbH & Co., Germany

Oil expellers are also manufactured by ECIRTEC, Brazil, and Ets Deklerck, Belgium.

16.0 Extruders

Extruding machines are used for a range of products, from cereals to nuts and meat. They basically consist of a tubular barrel, through which the food is passed while it is processed. Hot extruders are used for making cooked extruded products, such as snack foods, breakfast cereals and pet foods. The barrel contains a tight-fitting screw, similar to the expeller, which propels the foodstuff from one end of the barrel to the other as it turns and then is expelled through a small hole (the die) at the other end. The barrel is heated to process and cook the food.

EXTRUDERS
Used for pasta making. Capacity: up to 400 kg/hour
Power source: electric
Price code: 4
Agaram Industries, India

EXTRUDER: MODEL 600
Capable of cooking, expanding, sterilizing, dehydrating and texturing a wide range of products. By creating heat through friction, the dry extrusion process allows for high heat, short-cook time producing high quality feed and food. Capacity: 270–365 kg/hour
Power source: electric
Price code: 4
Alvan Blanch, UK

COCONUT MILK SCREW-TYPE EXTRACTION MACHINE
Capacity: 100 litres/hour
Power source: electric
Price code: 2
Charoenchai Company Ltd, Thailand

COOKER EXTRUDERS
A large range of co-rotating twin screw extruders with different outputs. Capacity: 20–15 000 tonnes/hour
Regis Machinery (Sales) Ltd, UK

17.0 Filling machines

The four basic methods for liquid filling are: vacuum filling, measured dosing, gravity filling and pressure filling. Filling by vacuum is the cleanest and most economical way to fill many products ranging from low to high viscosities, whereas gravity and pressure filling are best suited to the moderately rapid filling of low viscosity liquids such as fruit juices, which flow easily.

Powered

BOTTLE WASHING, FILLING, CAPPING/SEALING MACHINE
This combination machine can operate automatically or semi-automatically and is suitable for bottles from 0.5 to 1.5 litre. Washing operation is intermittent. Filling and capping/sealing operation is continuous. Low percentage of defective products. Capacity: 1000–2000 bottles/hour
Power source: electric
Price code: 4
ALFA Technology Transfer Centre, Vietnam

FILLING MACHINES
All kinds of automatic powder filling machines, automatic weighing machines, liquid and paste filling machines, tube filling and sealing machines etc.
Bombay Engineering Industry, India

FILLING AND PACKING MACHINES
A range of fillers for bottles, cups and cartons; sealing and capping systems.
Charles Wait (Process Plant) Ltd, UK

AUTOMATIC TUBE FILLING MACHINE
Used for filling and sealing tubes and bottles.
Capacity: 2000 per hour
Rank and Company, India

VACUUM FILLING MACHINE
Twin-head bottle-filling machine with vacuum pump and electric motor. There is an automatic overflow mechanism to catch excess liquid. Capacity: 5000 bottles per day
Power source: electric
Rank and Company, India

UNIVERSAL FILLING MACHINES
Filling machines for liquids and pastes; semi-automatic and fully automatic conveying systems; closing machines for inserts, cup to cup, screw caps, tablet presses.
Power source: manual/electric
J. Wick Filling Machines, Austria

FILLING MACHINE
This equipment has 4–12 filling heads and works by vacuum. Parts include: platform receiving bottles; plain chain conveyor; filling cabinet; filter; vacuum pump; automatic regulator with valve and timer. Suitable for plastic and glass bottles. Made of stainless steel and rustproof materials. Capacity: 1000–3500 bottles/hour
Power source: electric
Price code: 4
Technology & Equipment Development Centre (LIDUTA), Vietnam

Manual

FOOT OPERATED FILLING MACHINE
Used for filling tubes and bottles.
Power source: manual
Rank and Company, India

HAND OPERATED FILLING MACHINE
Used for filling tubes and bottles.
Power source: manual
Rank and Company, India

Filling machine equipment is also manufactured by Bombay Industrial Engineers, India (Form Fill Sealing Machines), Electra, France (Bag Filling Machine) and Pharmaconcept, India (Stainless Pipe Fitting).

17.1 Liquid fillers

Powered

CUP AND POUCH FILLERS
For liquids and pastes.
Capacity: 2000–20 000 units/hour
Power source: electric
Brasholanda SA – Equipamentos Industriais, Brazil

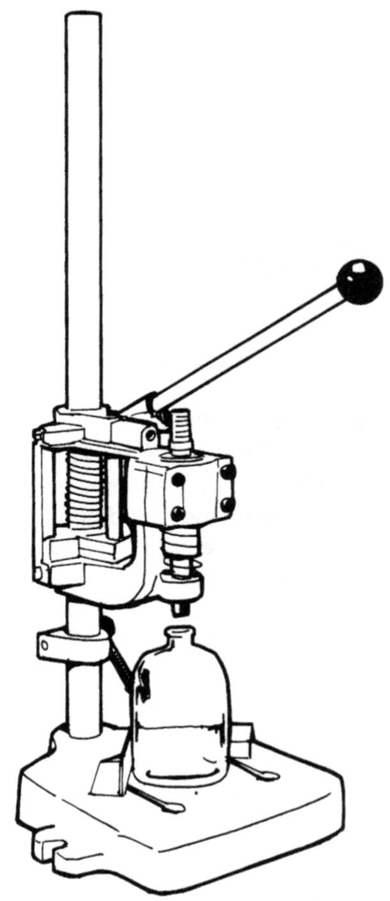

SEALING AND FILLING MACHINES
Semi-automatic machines for packing liquids such as milk, oil, ghee etc. in pillow packs. It is automatically operated. Capacity: 300 packs/hour
Power source: electric
Price code: 3
Dairy Udyog, India

VACUUM GLASS BOTTLE FILLER
Two and four heads. Capacity: 100–700 ml bottle
Eastend Engineering Company, India

THERMAL URN
Insulated to keep liquids boiling hot for use on a filling table. Capacity: 5–25 litres
Price code: 1
Gardners Corporation, India

BOTTLE FILLING MACHINE
5 ml – 1 litre bottles. Capacity: 1000+/hour
Pharmaco Machines, India

FILLING MACHINES
Suitable for filling jars using an electric motor or compressed air.
Power source: electric
Price code: 3
Swienty A/S, Denmark

Manual

BOTTLE/JAR FILLERS
Hand-operated stainless steel construction with adjustable fill volume. Accurately fills bottles, cans or cartons with a predetermined hot juice or pulp.
Power source: manual
Alvan Blanch, UK

PNEUMATIC LIQUID FILLER
Suitable for filling bottles with any food products in liquid form. No electric power is required. Capacity: up to 20 fills/minute
Power source: manual
Price code: 3
Autopack Machines Pvt Ltd, India

JUICE FILLING MACHINE
Power source: manual
Mark Industries (Pvt) Ltd, Bangladesh

DANA API MANA
A reliable and accurate machine for filling jars with honey, with a cut-off mechanism to prevent an extra drip. Capacity: 300–400 jars/hour
Power source: manual
Price code: 2
Swienty A/S, Denmark

Bottle-filling machines are also manufactured by Techno Equipments, India.

17.2 Solid fillers

PICKLE FILLING MACHINE
Price code: 2
Gardners Corporation, India

MANUAL SAUSAGE FILLERS
Capacity: 10–15 litres
Power source: manual
Price code: 3
Servifabri SA, Peru

SAUSAGE FILLER
Sausage filling machine made of iron with a stainless steel tube. Used to produce a range of sausages – hot dogs, chorizo and others. Available in two sizes. Capacity: 10 and 15 litres
Power source: manual
Price code: 2
Servifabri SA, Peru

FILLERS
Machinery for the sausage industry.
Capacity: 15–40 litres/hour
Talsabell SA, Spain

Pickle-filling machines are also manufactured by Geeta Food Engineering, India.

17.3 Paste fillers

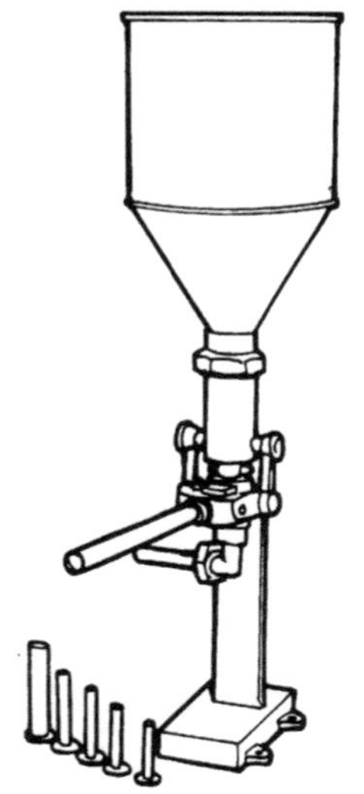

SEMI AUTOMATIC CREAM FILLER
Used for filling jars, bottles and containers with creams. Capacity: up to 2000 containers/hour
Power source: electric
Price code: 4
Autopack Machines Pvt Ltd, India

VOLUMETRIC FILLING APPARATUS
A small unit suitable for placing on a table for filling tubs, cups, jars etc. with semi-liquid or soft products. Capacity: maximum 500 g filling weight/stroke
Gerstenberg & Agger A/S, Denmark

VOLUMETRIC FILLING MACHINE
For filling buckets, cans, cartons etc. with soft margarine. Capacity: 0–5 kg filling weight/stroke
Gerstenberg & Agger A/S, Denmark

17.4 Powder fillers

TEA-BAG FILLING MACHINE
Part of a whole line that is used to process and pack tea and other dried goods. The delivery dose can be set from 2 to 10 g per dose. The machine can be used for a range of products, including tea, coffee and herbal teas. It has an integral electric counter and a range of other accessories. The machine has a three phase 1.8 hp motor. Capacity: 70–120 bags per minute
Power source: electric
Industrias Technologicas Dinamicas SA, Peru

18.0 Filters, sieves and strainers

Strainers

A common operation which often causes problems is the sieving/straining of wet products. This can be carried out in a pulper/finisher machine or, on a small scale, hand strainers can be used. Hand strainers are effective provided they are made to vibrate, which prevents blockage of the mesh.

FILTER BAGS OF TERYLENE
Felt for filtering juices (also known as jelly bags).
Price code: 1
Gardners Corporation, India

STARCH WASHER-EXTRACTOR
This machine is used to extract the starch by washing the pulp and sifting in water.
Capacity: 0.4–3 tonnes/hour
Gauthier, France

Filter equipment is also manufactured by Pharmaconcept, India (Stainless Steel Filter Housing).

18.1 Filter presses

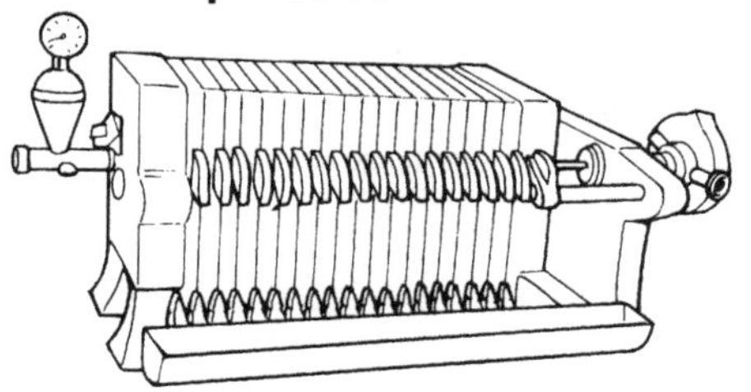

FILTER PRESS
Used for the filtration of oil from all kinds of oil-bearing seeds. Capacity: 10 kg/hour
Azad Engineering Company, India

JUICE FILTER UNIT
This machine consists of a vertical drum with a perforated sieve.
Price code: 1
Gardners Corporation, India

FILTRATION UNIT
Capacity: 0.5–2 tonnes/hour of fresh fruit
Gauthier, France

FILTER PRESS
This is suitable for the filtration of oil. It gives transparent, very clean oil without any solid particles. Capacity: 100–200 litres/hour
Price code: 2
Tinytech Plants, India

Filter presses are also manufactured by Pharmaconcept, India (Stainless Steel Filter Press).

18.2 Sieves

FLOUR SIFTER
Used for the grading of flour into various sizes. Rotary cylinder with conveniently replaceable woven mesh screens. Rotor with brushes and support frame with bagging outlets. Capacity: 20–600 kg/hour
Power source: manual/electric
Price code: 4
Alvan Blanch, UK

FRUIT PULPER/SIEVER
Used for the extraction of juice or pulp from fruit. Complete with feed chute, removable stainless steel perforated screen, rotary paddle with blades and collecting tray below.
Power source: electric
Price code: 4
Alvan Blanch, UK

FLOUR SIFTER
Designed for aerating and sifting flour.
Capacity: 2–5 tonnes/hour
Power source: diesel/electric
Price code: 4
Baker Enterprises, India

FLOUR SIFTER MACHINE
Power source: electric
Bijoy Engineers, India

FLOUR SIEVING MACHINE
Has an overhead hopper and discharge chute, complete with a floor stand.
Power source: electric
Price code: 2
Gardners Corporation, India

STAINLESS STEEL GAUZE SIEVE
Sieves have various mesh sizes and wooden sides.
Price code: 1
Gardners Corporation, India

ROTARY DRUM SIEVE
This machine is suitable for use as a pre-cleaner. Rotary brushes keep the sieve clean. The aspiration system keeps the working dust-free. Special rotary sieves are available for grading functions. Capacity: 100–2000 kg/hour
Power source: battery/electric
Price code: 3
Goldin (India) Equipment (Pvt) Ltd, India

'KEK' CENTRIFUGAL SIFTER
Power source: electric
Kemutec Group Ltd, UK

SIFTER
Portable and fitted with electric motor. Two different models of 20 and 30 inch diameter.
Power source: electric
Rank and Company, India

SIEVES
Automatic sieves for fruit juices and fruit pulp processing. This equipment is suitable for home or group use. Capacity: 40–150 kg/hour
Robot Coupe, France

18.3 Strainers

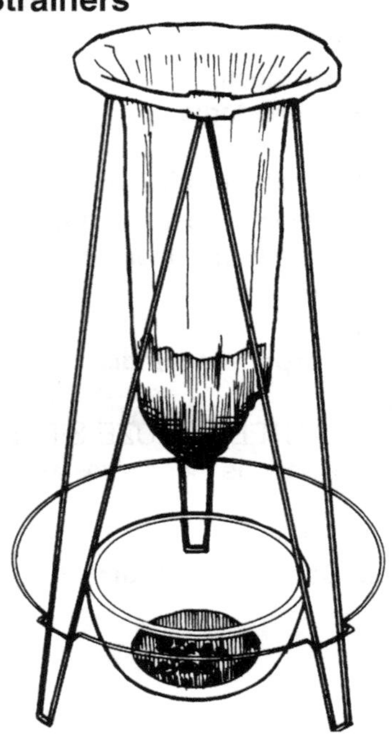

SOUP STRAINER
A very handy and fast device for pulping juices and tomatoes using a revolving paddle.
Power source: electric
Price code: 1
Gardners Corporation, India

CONICAL STRAINER
Strainer with nylon mesh. Very simple and very effective with honey and other liquids.
Power source: manual
Price code: 1
Swienty A/S, Denmark

19.0 Flaking and splitting equipment

Flaking and splitting machines are used for both cereals and pulses, to split pulses into dhal and to flake cereals. Flaking equipment is made from rollers that process cereal into elongated flakes.

SOYBEAN FLAKING MACHINE
This portable machine is used for both cereals and pulses. It consists of a hopper, hollow rollers, gear transmission system and a support frame. The rollers are fabricated from mild steel pipe and are surface plated with nickel. The power requirements are a 0.75 kW (1 hp) single phase electric motor through a belt drive to the central roller.
Capacity: 20 kg/hour
Power source: electric
Central Institute of Agricultural Engineering, India

20.0 Grating equipment

Grating equipment, both powered and manually operated, is available for a range of commodities.

COCONUT KERNEL GRATER
Motorized rotary grater.
Power source: electric/battery
Almeda Food Machinery Corporation, The Philippines

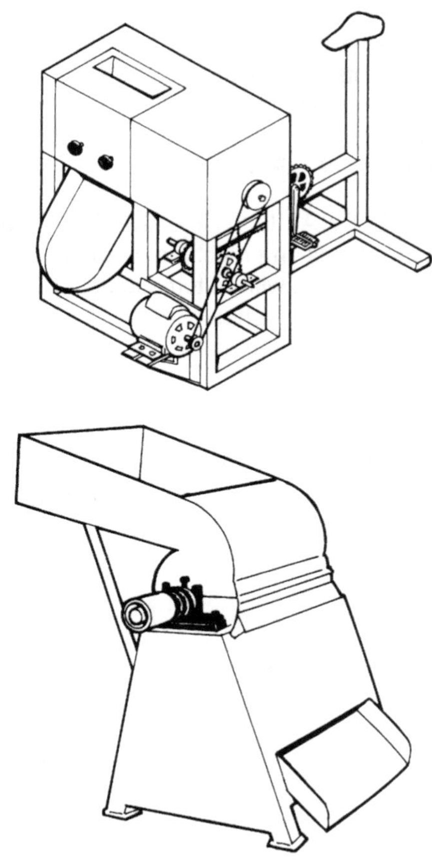

VILLAGE COCONUT SHREDDING MACHINE
Capacity: 60 kg/hour
Power source: electric
Price code: 1
Bay Hiap Thai Company Ltd, Thailand

ED500 CRUMBLING–DEFIBERING MACHINE
This machine is used to crumble the squeezed cassava pulp, by a grating process, and to eliminate the central fibres by sifting. Capacity: 0.5–3 tonnes/hour of fresh roots
Power source: electric
Gauthier, France

GRATERS
Power source: electric/manual
Natural Technik, Indonesia

COCONUT KERNEL GRATERS
Power source: manual/electric
Price code: 1
Odiris Engineering Company, Sri Lanka

COCONUT KERNEL GRATER
Power source: electric/manual
Price code: 1
Silva Industries, Sri Lanka

CASSAVA GRATER/CHIPPER
Cuts the cassava into small pieces to make the drying process quicker. Capacity: 200 kg/hour
Power source: manual
Uganda Small Scale Industries Association, Uganda.

21.0 Homogenizers

Homogenizers are used to form a stable emulsion from two immiscible liquids, usually oil and water. Homogenization causes the breakdown and dispersal of fat globules to form a homogenous liquid. In addition, it changes the functional properties or eating quality of foods, although it has no effect on the nutritional value or shelf life. Products that are homogenized include dairy products such as milk, ice cream, butter and margarine. An emulsifying agent such as lecithin may be added to assist the process. All homogenizers work on one of three basic principles: pressure, rapid mixing or ultrasonic vibration. The second type is more likely to be suitable for small-scale operations, but all types of equipment are expensive to buy and maintain. Many different types of equipment are available, ranging from simple hand-operated machines to larger, powered homogenizers with greater throughputs.

FT9 PRESSURE HOMOGENIZER
This machine is particularly useful for disintegrating fat globules in milk, ice cream and UHT dairy products. It can also be used for salad creams and sauces. Capacity: 45 litres/hour
Power source: electric
Armfield Ltd, UK

HOMOGENIZER
Capacity: 500 litres/batch
Power source: electric
Mark Industries (Pvt) Ltd, Bangladesh

PIVOTING LIQUIDIZER – SPECIAL
Made of stainless steel to allow for easy emptying. Comes with two sets of blades to make a smooth paste and for very fine grinding. Capacity: 20 and 25 litres
Power source: electric
Price code: 3
Servifabri SA, Peru

22.0 Ice cream equipment

A range of equipment, including mixers, moulds, liquidizers and pasteurizers that is used for making ice cream. Fully integrated ice cream makers and freezers are also available.

ICE CREAM MAKING EQUIPMENT
A range of products for the ice cream making process including pasteurizers, freezers, mixers, fillers and moulds.
Charles Wait (Process Plant) Ltd, UK

ICE CREAM MAKING EQUIPMENT
Power source: electric
RPM Engineers (India) Ltd, India

PNEUMATIC ICE CREAM DOSING MACHINE
Capacity: 18 litres
Power source: electric
Price code: 3
Servifabri SA, Peru

SOFT ICE CREAM MACHINE
Supplied with or without refrigeration unit and has a 3 hp motor. Capacity: 20 kg
Power source: battery/electric
Price code: 4
Servifabri SA, Peru

HORIZONTAL ICE CREAM MACHINE
Ice cream making machine, made of stainless steel, with automatic doser.
Capacity: 80 litres/hour
Power source: electric
Price code: 3
Servifabri SA, Peru

ICE CREAM FILLER
Ice cream machine for filling cartons or containers. Capacity: 18 litres
Power source: electric
Price code: 4
Servifabri SA, Peru

22.1 Freezers

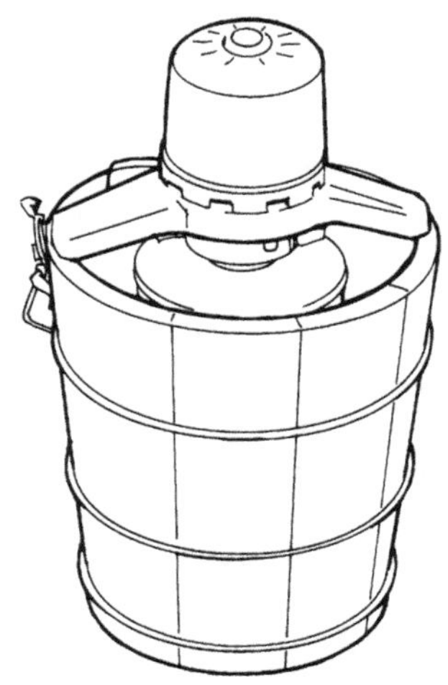

FT25A CONTINUOUS ICE CREAM FREEZER
Fully self-contained unit in mobile cabinet. The scraped surface unit is easily dismantled for cleaning and maintenance. Rapid cooling of viscous systems; ice cream freezing and air incorporation; iced desserts and yoghurts.
Capacity: 10 litres/hour
Power source: electric
Armfield Ltd, UK

ICE CONFECTIONERY FREEZING MACHINE
Power source: electric
Price code: 1
Bay Hiap Thai Company Ltd, Thailand

23.0 Mills and grinders

A wide selection of mills and grinders, suitable for grinding cereals, legumes, herbs, spices, nuts, oil-seeds, coffee and sugar is available in a range of sizes. Manually operated and powered versions of most mills are available.

Powered

COFFEE GRINDER KF1800
Used for the grinding of coffee after roasting. This machine has a large 30 kg hopper and adjustable outlet pipe. An air cooling system eliminates the need for traditional water-cooling of the machine.
Power source: electric
Price code: 4
Alvan Blanch, UK

WET MILLING MACHINE
Used for milling rice. Capacity can be adjusted according to customer needs.
Capacity: 20 kg/hour
Power source: electric
Price code: 1
Anh Tuan Mechanical Cooperative, Vietnam

FINE GRINDER
This machine can be used to separate creamed desiccated coconut into virgin coconut oil and a fine aromatic desiccated coconut, as well as other fruits and vegetables.
Power source: electric
Arai Machinery Corporation, Japan

MILLING MACHINE
Power source: diesel
Atelier de Soudure Tout por Moulins, Burkina Faso

MILLING MACHINES
Power source: diesel/electric
Price code: 2
Atelier de Soudure Tout por Moulins, Burkina Faso

OIL MILL MACHINERY: OIL EXPELLERS, FILTER PRESSES AND SPARE PARTS
Extraction and filtration of oil. Capacity: varies from machine to machine
Power source: electric
Azad Engineering Company, India

COCONUT JELLY DISINTEGRATING MACHINE
Capacity: 250 kg/hour
Power source: electric
Price code: 2
Bay Hiap Thai Company Ltd, Thailand

SUGAR PULVERISING MACHINE
Power source: electric
Bijoy Engineers, India

MINI INTEGRATED PULSE (DAHL) MILL
The machine grades the product into fractions of unsplit de-husked legume, split whole legume, broken pieces and husk. It consists of a de-hulling unit and dahl separator and an aspirator. A vibrating sifter does grading and sorting. Capacity: 100–125 kg/hour
Power source: electric
Central Food Technological Research Institute, India

DHAL MILL
A low cost unit for de-husking and splitting of pulses. Capacity: 100 kg/hour
Power source: electric
Central Institute of Agricultural Engineering, India

GENERAL PURPOSE MILL
Power-operated equipment for pearling or de-hulling of coarse cereals and pulses. It removes bran from wheat and coarse grains, and husk from pulses. Suitable for small-scale pearling/polishing units. Capacity: 100–300 kg/hour
Power source: electric
Central Institute of Agricultural Engineering, India

LOW COST MULTIPURPOSE GRAIN MILL
Suitable for grinding wheat, sorghum, soybean, Bengal gram and coriander. Makes grits, flour, powder and splits. Offers precise control for particle size. Foundation is not required for installation. Capacity: 10–70 kg/hour
Power source: electric
Central Institute of Agricultural Engineering, India

CURRY PASTE GRINDING MACHINE
Capacity: 200 kg/hour
Power source: electric
Price code: 3
Charoenchai Company Ltd, Thailand

WET SOYBEAN GRINDING MACHINE
Capacity: 80 kg/hour
Power source: electric
Price code: 2
Charoenchai Company Ltd, Thailand

LAMILPA POWER MILL
This machine will grind wet or dry material. It is adjustable for coarse to fine and has a feed auger with 12 inch V-pulleys. Capacity: 150 kg/hour
Power source: electric
Price code: 2
C.S. Bell Co., USA

RICE MILL
Used for converting paddy to final rice.
Capacity: 250–2500 kg
Power source: diesel/electric
DIW Precision Engineering Works, India

MILLING MACHINE
Mills rice to flour with water for making ravioli and rice pancakes. Capacity: 40 kg/hour
Power source: electric
Price code: 1
Doan Binh Mechanical Cooperative, Vietnam

FRUIT MILLER
Capacity: 250–1000 kg/hour
Eastend Engineering Company, India

MILLING MACHINES
Power source: diesel
El Hadj Yaya Akewoula, Burkina Faso

GRINDER FOR GROUNDNUT PASTE
Power source: battery/electric
Electra, France

PALM NUT CRUSHER
Power source: battery/electric
Electra, France

MILLING MACHINES
Power source: diesel
Garage Outtara Bakari, Burkina Faso

FRUIT MILL
This machine is designed for stoneless fruits and vegetables. It has a feed opening cap with a hopper discharge chute. Capacity: 1000 kg/hour
Power source: electric
Price code: 2
Gardners Corporation, India

POWER GRINDER
Capacity: 5–10 kg/hour
Power source: electric
Price code: 1
Gardners Corporation, India

SUGAR GRINDING MACHINE
Capacity: 20–25 kg/day
Power source: electric
Price code: 2
Gardners Corporation, India

GRINDMASTER MODEL 3000
Commercial peanut grinding equipment designed for shop-counter use. Capacity: 0.5 kg peanuts in 30 seconds
Power source: electric
Price code: 3
Grindmaster Corporation, USA

MILLING MACHINE
Power source: battery
Guire Harouna (Etablissement), Burkina Faso

GRINDING MILLS
Machinery for processing maize. Capacity: 300–1200 kg/hour
Power source: diesel/electric
H.C. Bell & Son Engineers (Pvt) Ltd, Zimbabwe

GRAIN MILL
Power source: electric
Joint-Stock Company Radviliskis Machine Factory, Lithuania

GRINDING CRUSHING EQUIPMENT
Power source: diesel
Kabore Moussa et Frères (Etablissement), Burkina Faso

MS STAND (BATCH OPERATION)
Recommended for smaller equipment of lower throughput capacity, or where frequent changeover of product is a criterion.
Power source: electric
Kaps Engineers, India

VM AIR CLASSIFIERS
Desired particles are conveyed for discharge, while coarser particles settle down for collection or recycling. Capacity: 5–500 kg/hour
Kaps Engineers, India

VM PRE-CRUSHER
Designed as a primary crusher which can be used prior to fine grinding equipment. Lumps and irregularly shaped feed materials are crushed into a mixture of small lumps to coarse powder. Throughput capacity and size depend on the properties of material being crushed. Capacity: 25–5000 kg/hour
Power source: electric
Kaps Engineers, India

VM PULVERIZERS
Recommended for medium fine grinding of soft to semi-hard materials. Equipped with changeable grinding elements. Capacity: 100–3000 kg/hour
Power source: electric
Kaps Engineers, India

COFFEE BEAN GRINDING MACHINE
Capacity: 100 kg/hour
Power source: electric
Price code: 3
Kasetsart University, Thailand

'KEK' KIBBLERS
Stand-alone coarse grinding mills which are designed to accept lumps up to 150 mm diameter (and even larger rectangular shapes) and reduce them to 2 or 3 mm.
Capacity: 2–25 tonnes/hour
Power source: electric
Kemutec Group Ltd, UK

GROUNDNUT COARSE GRINDING MACHINE
Capacity: 275 kg/hour
Power source: electric
Price code: 3
Kunasin Machinery, Thailand

VICTORIA ELECTRICAL GRINDER
Body, base and parts are manufactured in cast iron with sanitary tinned finish. Hopper in tin-plated steel. Sealed gear box with four gears.
Power source: electric
Mecanicos Unidos, Colombia

VICTORIA GRAIN MILL – HIGH HOPPER

The grinding plates are adjustable for various degrees of grinding. To avoid any food contamination, the mill is coated with a thick plate of pure tin. The grain grinder can be easily dismantled after use for washing and cleaning.
Mecanicos Unidos, Colombia

MILLING MACHINERY

Power source: electric/diesel
Nacanabo (Etablissement), Burkina Faso

OIL MILL

Tiny oil mills in three models for multipurpose grains – sunflower, rapeseed, groundnut, cotton, shea nut, sesame and cocoa etc. Oil cake production from 7–12 per cent.
Capacity: 175–450 kg/hour
Olier Ingenierie, France

MILLING MACHINE

Power source: electric
Ouedroago Safiatou (Ets), Burkina Faso

MILLS

A range of mills capable of grinding material to 100–250 mesh particle size. Capacity: 5–1000 kg/hour
Power source: diesel/electric
Price code: 2
Premium Engineers Pvt Ltd, India

SINGLE PASS MINI RICE MILL

De-husks the paddy by rubber rolls in the upper chamber. Separates the husk, dust and immatured grain in the middle chamber by the powerful aspirator and polishes the brown rice, removing the bran at the lower chamber. The white polished rice is discharged through the outlet chute. Capacity: 400–850 kg/hour
Power source: electric
Rajan Universal Exports (Manufacturers) Pvt Ltd, India

VERTICAL STONE GRINDING MILL

The grinding is done by a pair of emery grinding stones. Capacity: 150–600 kg/hour
Power source: electric
Rajan Universal Exports (Manufacturers) Pvt Ltd, India

MULTI MILL

For high-speed granulation, pulverization etc. of wide range of wet and dry materials. This machine has 12 blades with knife and impact edges.
Rank and Company, India

MILLING MACHINE

This machine is a rice miller.
Power source: diesel/electric
Price code: 3
Ruang Thong Machinery Ltd, Thailand

MILLING MACHINES

Power source: diesel
Sempore Issa et Frères (Etablissement), Burkina Faso

CRUSHING, MILLING, GRINDING MACHINERY

Power source: diesel
Société de Production et d'Exploitation de Matérial de Traitement du Karite, Burkina Faso

MINI MILLS

Rice, dhal, grain and wheat mills.
Sri Murugan Industries, India

MILLING MACHINES

Power source: electric/diesel
Tabsoba Salifou et Frères (Etablissement), Burkina Faso

GRINDING MACHINES

A range of free-standing models available with electric (4–12 kW motors) or diesel engines. Capacity: 250–2000 kg/hour
Power source: electric/diesel
Udaya Industries, Sri Lanka

SHEA NUT CRUSHING MACHINE
Power source: electric
WA ITTU, Ghana

STONE GRINDING MILLS
For fine or coarse grinding of crops.
Capacity: 15–1000 kg/hour
Alvan Blanch, UK

OIL MILL MACHINERY
Used for the extraction and filtration of all kinds of oil-bearing seeds. Capacity: 10 kg/hour
Azad Engineering Company, India

NO. 60 POWER GRIST MILL
Suitable for use on large and small farms. Adjustable for grinding texture. Includes hopper, feed regulator slide, coarse and fine grinding burrs and 12 inch diameter pulley.
Capacity: 150 kg/hour
Power source: electric/manual
Price code: 2
C.S. Bell Co., USA

MILLING MACHINE
Used for crushing palm kernels, groundnut seeds and moringa seeds to extract oil.
Power source: electric/manual
Price code: 3
Nigerian Oil Mills Ltd, Nigeria

Manual

NO. 2 HAND GRIST MILL
Adjustable for coarse to fine grinding with counterbalanced handle crank. Holds five cups of grain. Cone-shaped burrs are made of special alloy for longer life. Capacity: 5 kg/hour
Power source: manual
Price code: 1
C.S. Bell Co., USA

'MFM' DOMESTIC GRAIN GRINDING MILL
Designed for home use. This machine has a simple adjustment for fine or coarse meal grinding. Capacity: 7 kg/hour
Power source: manual
G. North & Son (Pvt) Ltd, Zimbabwe

'NATIONAL' GRAIN GRINDING MILL
Specifically designed for the smallholder.
Capacity: 30 kg/hour
Power source: manual
G. North & Son (Pvt) Ltd, Zimbabwe

GRAIN MILL
Adjusts from powder fine to flaky coarse.
Power source: manual
Price code: 1
Lehman Hardware & Appliances, USA

PEANUT BUTTER GRINDER
Hand-powered food mill to make peanut butter.
Power source: manual/electric
Price code: 1/2
Lehman Hardware & Appliances, USA

MANUAL MILL
Power source: manual
Samap, France

Mills and grinders are also manufactured by AMI Engineering, India (Semi-Automatic Mini Dal Mills); Electra, France (Cassava Grinder); Outils Pour Les Communautes, Cameroon (Cereals Mill, Cassava Grinder); Sri Rajalakshmi Commercial Kitchen Equipment, India (Flour Mill, Wet Grinders); WA ITTU, Ghana (Grain Milling Machines).

23.1 Plate mills

Plate mills are usually limited to about 7 kW and are based on the stone mill or quern. In the modern plate mill, two chilled iron plates are mounted on a horizontal axis so that one of the plates rotates and the grain is ground between them. The pressure between them governs the fineness of the product and is adjusted by a hand screw. The grain is usually coarsely cracked in the feed screw to the centre of the plates. Grooves in the plates decrease in depth outwards towards the periphery so that the grain is ground progressively finer until it emerges at the outer edge and falls by gravity into a sack or bowl.

Plate mills are also effective at grinding

wet products such as wetted maize, tomatoes, peppers and spices. Manual versions of the plate mill are available, but the throughput is only about 1–2 kg per hour.

PLATE GRINDING MILLS
Suitable for wet and dry grinding of grains and other crops. Supplied with two types of plates for fine and coarse milling. Capacity: 15–1000 kg/hour
Power source: manual/diesel/electric
Alvan Blanch, UK

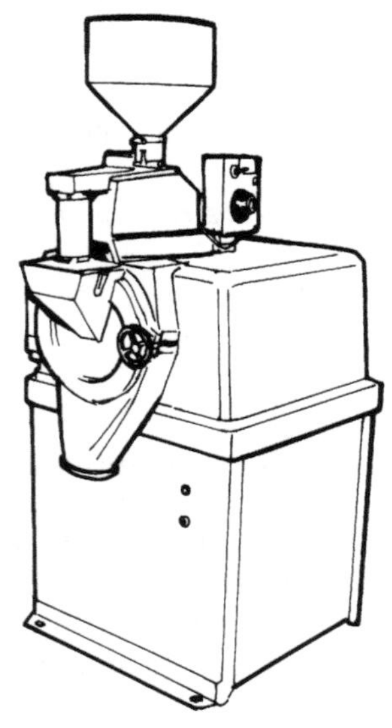

DISC GRINDER
This machine reduces a crushed product to a fine particle size. It has two 20.3 cm diameter grinding discs: one fixed and one rotating. The material falls from a feed chute through a central hole in the fixed disc into the grinding chamber, and the ground material is discharged from a chute at the front of the machine. Capacity: 50 kg/hour
Power source: diesel/electric
Price code: 4
Pascall Engineering Company, UK

23.2 Roller mills

A roller mill flattens grain rather than grinding it. This type of mill is most useful for producing animal food as ruminants require crushed grain. A machine with a 3 kW engine is capable of crushing up to half a tonne of barley per hour.

ROLLER MILLS
A range of sizes of roller mills are available.
Capacity: 500–1000 kg/hour
Power source: electric
Alvan Blanch, UK

RICE DOUGH ROLLER
This machine is used for rolling rice dough. The capacity can be adjusted according to customer needs. Capacity: 100 kg/hour
Power source: electric
Price code: 1
Duc Huan Mechanical Cooperative, Vietnam

TRIPLE ROLL MILL
For grinding paste.
Rank and Company, India

CEREAL ROLLER MILL
Suitable for grinding a range of cereals including barley, wheat, quinoa, kiwicha and beans. Capacity: 100–1000 kg/hour
Vulcano Technologia Aplicada Eirl, Peru

Roller mills are also manufactured by Electra, France (Roller Mill); Samap, France (Cereal Roller Mill); and Sree Manjunatha Roller Flour Mills (P) Ltd, India (Roller Flour Mill).

23.3 Hammer mills

Hammer mills work on the principle that most materials will crush or pulverize upon impact using a simple three-step operation:

1. Material is fed into the mill's chamber, typically by gravity.
2. The material is struck by ganged hammers (rectangular pieces of hardened steel) which are attached to a shaft which

rotates at high speed inside the chamber. The material is crushed or shattered by the repeated hammer impacts, collisions with the walls of the grinding chamber, as well as particle on particle impacts.

3. Perforated metal screens or bar grates covering the discharge opening of the mill retain coarse materials for further grinding while allowing the properly sized materials to pass as finished product.

Varying the screen size, shaft speed, or hammer configuration can dramatically alter the finished size of the product being ground. Faster speed, a smaller screen and more hammers result in a finer end product. Each component can be changed individually or in any combination to produce the precise grind required.

Most grain crops can be ground in a hammer mill.

SUPERMILLS
These machines operate on a gravity feed and gravity discharge basis. Capacity: 80–400 kg/hour
Power source: electric
Price code: 3
Alvan Blanch, UK

BRITON GRINDER
A range of hammer mills for the production of all kinds of feeds and cereal grinding.
Power source: electric
Christy, UK

BRITON AGRICULTURAL GRINDER
Feed is direct from grain bin into suction hopper. Automatic cut out fitted in hopper to stop grinder when bin is empty.
Power source: electric
Christy, UK

DISCHARGE HAMMER MILL
Includes basic mill, feed regulating table, ¼ inch screen, motor mounting bracket, V-belts and pulleys.
Power source: electric
Price code: 3
C.S. Bell Co., USA

HAMMER MILL
Capacity: 100–400 kg/hour
Power source: diesel
DISEG – Peru

HAMMER MILL
Delivery and discharge chutes. Capacity: 200–650 kg/hour
Power source: electric
Price code: 2
Gardners Corporation, India

HAMMER MILL
6 hp/1700 rpm motor with 24 stainless steel hammers. Capacity: 100–150 kg/hour
Power source: electric
Price code: 3
Industrias Technologicas Dinamicas SA, Peru

HAMMER MILL
15 hp/1700 rpm motor with 48 stainless steel hammers. Capacity: 200–250 kg/hour
Power source: electric
Price code: 3
Industrias Technologicas Dinamicas SA, Peru

HAMMER MILL
Stainless steel mill with 48 blades that can be reversed four times. Has a 15 hp motor.
Capacity: 200–350 kg/hour
Power source: electric
Price code: 3
Industrias Technologicas Dinamicas SA, Peru.

VM AIR SWEPT MILLS
Recommended for micro-fine pulverizing of soft to medium hard fibrous and slightly moist materials. Capacity: 250–800 kg/hour
Kaps Engineers, India

VM MIKRO PULVERIZERS
A built-in water circulation jacket retains grinding temperature lower than otherwise. The design provides for admitting extra air during grinding to facilitate cool operation. Soft to medium hard materials are quickly pulverized to fine powder. Capacity: 25–40 kg/hour
Kaps Engineers, India

HAMMER MILL
Capacity: 750 kg/hour
Power source: electric
Price code: 2
Kundasala Engineers, Sri Lanka

MIRACLE MILL
Precision reduction up to 60 mesh. This machine is a three in one unit which pulverizes, sieves and collects. It cool-grinds anything grindable to a specified degree of fineness in one operation using easily changeable screens.
Capacity: 50–6000 kg/hour
Power source: diesel/electric
Price code: 3
Premium Engineers Pvt Ltd, India

HAMMER MILL
Capacity: 75–1000 kg/hour
Power source: electric
Rajan Universal Exports (Manufacturers) Pvt Ltd, India

POPULAR HAMMER MILL
High output gravity flow mill; large feed hopper; simple feed adjustment; full size swinging hammers; and bagging off chute.
Capacity: 250–1200 kg/hour
Power source: electric
Scotmec (Ayr) Ltd, UK

TROPICAL HAMMER MILL
Fan-type elevating hammer-mill for general use with suction feed unit and vibrator feeder available. Capacity: 500–2000 kg/hour
Power source: diesel/electric
Price code: 3
Scotmec (Ayr) Ltd, UK

UNIVERSAL HAMMERMILLS
Fan-type elevating mills for grinding grain, straw, oil cake etc. with bulk feed top; double bagging cyclone; combined chute-hopper and reversible swinging hammers.
Scotmec (Ayr) Ltd, UK

HAMMER MILL
Available in a range of sizes, this machine is useful for grinding a range of cereals and pulses. Capacity: 50–2000 kg/hour
Vulcano Technologia Aplicada Eirl, Peru

Hammer mills are also manufactured by Electra, France.

23.4 Colloid mills

Colloid mills work by hydraulic shear to break up particles into crystal size. A limitless variety of materials can be processed in a colloid mill.

HOMOGENIZER (COLLOID MILL)
The purpose of the homogenizer is to form an emulsion and to break down aggregates and disperse the ultimate particles in liquids. Powered by 5/10/15 hp motor. Capacity: 100–800 litres/hour
Power source: electric
Capitol Engineering Works, India

HOME COLLOID MILL
Capacity: 10–1200 kg/hour
Cip Machineries Pvt Ltd, India

24.0 Mixers

A range of mixers and blenders for small- and medium-scale food processing is available.

Planetary mixers have two or three vertical mixing blades that rotate together. They also have a scraping blade that prevents material from adhering to the side wall of the vessel and ensures more even mixing. Some models have a centralized high-speed dispersion blade that enhances product homogeneity.

Powered

MIXERS AND BLENDERS
A range of mixers and blenders for small- and large-scale food processing.
Charles Wait (Process Plant) Ltd, UK

POWDER MIXER
A stainless steel, single phase, screw mixer that causes two-way mixing. Capacity: 50–100 litres
Power source: electric
DISEG (Diseno Industrial y Servicios Generales), Peru

MIXER
Double jacketed pan and horizontal mixer.
FAINSA (Fabricantes En Acero Inoxidable SA), Peru

MIXER, BLENDER AND SLICER
Performs the following functions: the mixing and blending of liquids; production of finger chips of potatoes and hard stoneless fruits; the slicing and shredding of potatoes, cabbage, carrots, apples etc.; the crushing of ice; kneading of flour, mince meat and fruits; the whisking and beating of eggs.
Power source: electric
Price code: 1
Gardners Corporation, India

PICKLE MIXING MACHINE
Capacity: 200 kg/hour
Power source: electric
Price code: 2
Gardners Corporation, India

BATCH MIXER – TYPE FKM
This machine is in a heated, encapsulated construction for pharmaceutical application.
Gebruder Lodige Maschinenbau GmbH, Germany

BATCH MIXER – TYPE FKM-A
This machine has withdrawable mixing elements. Capacity: 300–3000 litres total volume
Gebruder Lodige Maschinenbau GmbH, Germany

CONTINUOUS MIXER – TYPE CB20-CB100
High speed mixer. Capacity: 100–200 kg/hour
Power source: electric
Gebruder Lodige Maschinenbau GmbH, Germany

CONTINUOUS PLOUGHSHARE MIXER – TYPE KM
Capacity: 150–15 000 litres total volume
Gebruder Lodige Maschinenbau GmbH, Germany

HEATING/COOLING MIXER COMBINATIONS – TYPE TSHK/KGU
Capacity: 100–4000 litres total volume
Gebruder Lodige Maschinenbau GmbH, Germany

HIGH SPEED MIXER – TYPE TSHK
Capacity: 100–2000 litres total volume
Gebruder Lodige Maschinenbau GmbH, Germany

PLOUGHSHARE MIXER – FKM RANGE
Suitable for batch operation.
Gebruder Lodige Maschinenbau GmbH, Germany

UNIVERSAL MIXER – TYPE KUM
Capacity: 100–1000 litres total volume
Gebruder Lodige Maschinenbau GmbH, Germany

HORIZONTAL MIXER
This stainless steel mixer has a 3.5 hp motor and operates at 105 rpm. Capacity: 600 kg/hour
Power source: electric
Price code: 3
Industrias Technologicas Dinamicas SA, Peru

'GARDNER' MIXERS AND BLENDERS
A full range of easy to clean small batch machines, gentle ribbon mixers, high-speed high-intensity plough share mixers and double cone blenders.
Power source: electric
Kemutec Group Ltd, UK

HORIZONTAL MIXER
Using this mixer, the mixing cycle is fast, and dry or wet mixing can be done. Discharge opening is provided at the bottom to help in rapid discharge. These mixers are ideal where liquid molasses up to 10 per cent are required. Mixing is completed within 7 to 10 minutes. Capacity: 125–1000 kg/hour
Power source: electric/diesel
Premium Engineers Pvt Ltd, India

POWDER MASS MIX
Suitable for mixing wet and dry powders, with mechanical tilting arrangements.
Power source: electric
Rank and Company, India

HORIZONTAL FEED MIXERS
Suitable for mixing all feeds. Slow-moving soft mix action eliminates product damage and degradation.
Scotmec (Ayr) Ltd, UK

MIXERS
A range of machinery for the sausage industry. Capacity: 35–150 litres/hour
Power source: electric
Talsabell SA, Spain

MIXING VESSELS AND MACHINES FOR LIQUIDS, PASTES AND POWDERS
Power source: electric
Technical Sales International, UK

THE TATHAM MIXER
High speed and gentle mixer for dry solids, powders, flakes, granules and solids with liquid additions. Capacity: 20–5000 litres
Power source: electric
Winkworth Machinery Ltd, UK

Manual

STIRRING SCREW
Used for honey processing.
Power source: manual
Price code: 1
Swienty A/S, Denmark

Mixers are also manufactured by Alven Foodpro Systems (P) Ltd, India, and Sri Murugan Industries, India

24.1 Dough mixers

DOUGH MIXING AND BEATING MACHINES
Machines come with grinder attachments.
Power source: electric
AMI Engineering, India

FOOD MIXING MACHINE
Powder mixer. Capacity: 60 kg/hour
Power source: electric
Price code: 2
Amorn Loharyon, Thailand

DOUGH MIXER
Two speed with a 2.4/3.6 hp motor. Capacity: 25 kg
Power source: electric
ANLIN, Peru

DOUGH KNEADER
Designed to mix flour with other ingredients. The kneader is supplied with one mixing pan mounted on a wheeled trolley enabling free movement from one place to another. Motor 3 hp. Capacity: 50–170 kg
Power source: diesel/electric
Price code: 4
Baker Enterprises, India

FLOUR KNEADING MACHINE
Price code: 2
Gardners Corporation, India

DOUGH MIXER
Capacity: 25 kg/hour
Power source: electric
Industria Peruana Comercializadora Techno Pan Equipamiento Integral, Peru

FLOUR DOUGH MIXING MACHINE
Capacity: 50 kg/hour
Power source: electric
Price code: 2
Krungthep Chanya, Thailand

Dough mixers are also manufactured by Samap, France (Kneading Machine), and Sri Rajalakshmi Commercial Kitchen Equipment, India (Dough Kneaders).

24.2 Cake mixers

PLANETARY MIXER
This machine is designed to suit large-, medium- and small-scale bakeries for uniform mixing of cake batters and small quantities of bread and pie doughs. Attachable items – three stainless steel beaters and a stainless steel pan. Three speeds are available for kneading, mixing or whipping. Capacity: 25–30 litres
Power source: diesel/electric
Price code: 4
Baker Enterprises, India

PLANETARY CAKE MIXING MACHINE
Stainless steel bowl with three-speed pulley system. Capacity: 25–40 kg/hour
Power source: electric
Bijoy Engineers, India

MIXERS
High-speed mixers and kneaders.
Bombay Industrial Engineers, India

SEALED MIXER
Industrial equipment. Capacity: 250 litres
FAMET IUGS SRL, Peru

CAKE MIXING MACHINE
Planetary mixer suitable for egg beating and dough mixing with floor stand.
Power source: electric
Price code: 2
Gardners Corporation, India

COOLING MIXER – TYPE KGU
Capacity: 250–4000 litres total volume
Gebruder Lodige Maschinenbau GmbH, Germany

GRIDLAP CAKE MIXERS
Capacity: 251–319 litres total volume
Gebruder Lodige Maschinenbau GmbH, Germany

Z-BLADE DUPLEX MIXERS
Capacity: 1726 litres total volume
Gebruder Lodige Maschinenbau GmbH, Germany

Z-ARM POWDERED FOOD MIXER
Capacity: 200 kg/hour
Power source: electric
Price code: 2
KSL Engineering, Thailand

DRUM MIXER
For quick granule mixing and lubricating.
Capacity: 200 litres total volume
Rank and Company, India

PLANETARY MIXER
Most versatile and efficient in mixing ointment, pastes, emulsion, powder etc. with three speed gears. Capacity: 10–1000 litres
Rank and Company, India

PLANETARY MIXER
Capacity: 20–30 litres
Power source: electric
Price code: 3
Servifabri SA, Peru

PLANETARY MIXER
Planetary mixer with a stainless steel bowl, suitable for a range of foods. It is useful for whipping, beating and kneading. The mixer uses a planetary action for uniform mixing. It is available in two sizes, requiring a 1 or 1.5 hp motor. Capacity: 20 or 30 litres
Power source: electric
Price code: 3
Servifabri SA, Peru

HIGH SPEED 'RT' MIXERS
Heavy duty jacketed model with loading hopper, high-speed refiner and safety gate around the outlet. Capacity: 10–7000 litres
Power source: electric
Winkworth Machinery Ltd, UK

HOBART MIXERS
Three speeds. Detachable stainless steel bowl with aluminium beater, hook and whisk.
Power source: electric
Winkworth Machinery Ltd, UK

STAINLESS STEEL 'RT' MIXERS
For mixing all types of pastes. Capacity: 10–7000 litres
Power source: electric
Winkworth Machinery Ltd, UK

Cake mixers are also manufactured by: Premium Engineers Pvt Ltd, India (Vertical Mixer), and Technoheat Ovens and Furnaces Pvt Ltd, India.

24.3 Liquidizers

LIQUIDISER
This is a tilting 20 litre industrial liquidizer with a 1 hp motor.
Power source: electric
DISEG (Diseno Industrial y Servicios Generales), Peru

LIQUIDIZER
A semi-industrial 5 litre liquidizer.
Power source: electric
DISEG (Diseno Industrial y Servicios Generales), Peru

ELECTRIC LIQUIDIZER
Suitable for mixing and liquefying of fruits, vegetables and flour kneading.
Power source: electric
Price code: 1
Gardners Corporation, India

WET MIXER – TYPE NOSHK
Capacity: 75–10 000 litres total volume
Power source: electric
Gebruder Lodige Maschinenbau GmbH, Germany

EMULSIFIER
Used for suspensions and emulsions etc. Complete with motor and stirring head. Special arrangements for mechanical lifting and lowering down.
Rank and Company, India

ROTO CUBE
Used for homogenous mixing of dry powder, syrup and lubricants or granules. Vessels and blades rotate at different speeds. Capacity: 50–200 kg
Rank and Company, India

STIRRER
High-speed stirrer, complete with motor for mixing syrup, oils and emulsions. Capacity: 200–1000 litres
Rank and Company, India

TABLE LIQUIDIZERS
Four-blade table liquidizers with 1 hp motors and 4500 rpm. Capacity: 3–5 litres
Power source: electric
Price code: 2
Servifabri SA, Peru

TILTABLE LIQUIDIZERS
These machines have four blades and between 0.75 and 2 hp motors with 3500 rpm. Capacity: 10–25 litres
Power source: battery/electric
Price code: 2/3
Servifabri SA, Peru

TILTABLE LIQUIDIZERS
Special high-speed four-blade tiltable liquidizers with 1.5–2 hp motor and 4500 rpm. Capacity: 20–25 litres
Power source: electric
Price code: 3
Servifabri SA, Peru

PIVOTING LIQUIDIZER
Liquidizer made of stainless steel that pivots to allow for easy emptying. Available in a range of sizes with motors ranging from 0.75 to 2 hp. Capacity: 10–25 litres
Power source: electric
Price code: 3
Servifabri SA, Peru

24.4 Blenders

MIXER, BLENDER AND SLICER
Performs the following functions: the mixing and blending of liquids; production of finger chips of potatoes and hard stoneless fruits; the slicing and shredding of potatoes, cabbage, carrots, apples etc.; the crushing of ice; kneading of flour, mince meat and fruits; the whisking and beating of eggs.
Power source: electric
Price code: 1
Gardners Corporation, India

RIBBON-BLENDER
Capacity: 45 litres total volume
Gebruder Lodige Maschinenbau GmbH, Germany

BLENDER
Capacity: 500 kg/hour
Power source: electric
Price code: 3
Premium Engineers Pvt Ltd, India

DOUBLE CONE BLENDER
Most efficient and versatile for mixing of light powders homogeneously. Capacity: 5–500 kg
Rank and Company, India

HORIZONTAL BLENDER
Horizontal ribbon blender.
Power source: electric
Vulcano Technologia Aplicada Eirl, Peru

DOUBLE CONE BLENDERS AND BLENDER DRYERS
Meets the specialized requirements of gentle mixing. Used for powders, granules, flakes etc. Capacity: 1–8000 litres
Winkworth Machinery Ltd, UK

Blenders are also manufactured by Sri Venkateswara Industries, India.

25.0 Ovens

Ovens, stoves and integrated baking machines in a range of styles and sizes are available.

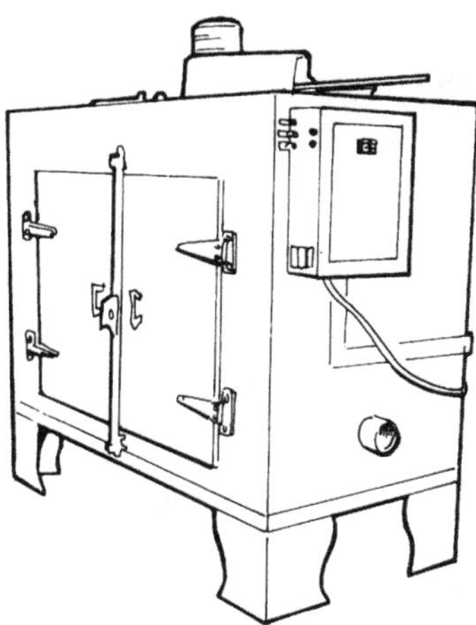

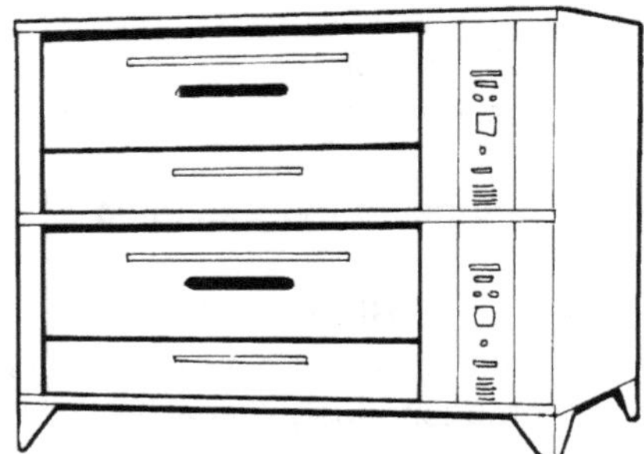

INDUSTRIAL OVENS
All types and any size of diesel-fired automatic ovens for biscuits, bread and rusks.
Power source: diesel
Admir Enterprises, India

DOUBLE MAXI
Cooking, smoking and drying system. Designed for small batch processing of meat, fish and cheese etc. Three smoke box drawers are fitted at the left-hand side of the unit in which sawdust is placed and manually ignited. Capacity: 127 kg/hour
Power source: electric
Price code: 4
AFOS Ltd, UK

MAXI
Cooking, smoking and drying system. Designed for small batch processing of meat, fish and cheese etc. Three smoke box drawers are fitted at the left-hand side of the unit in which sawdust is placed and manually ignited. Capacity: 50 kg/hour
Power source: electric
Price code: 4
AFOS Ltd, UK

MINI
Smoking, cooking and drying system. Designed for small batch processing of meat, fish and cheese etc. Three smoke box drawers are fitted at the left-hand side of the unit in which sawdust is placed and manually ignited. Capacity: 25 kg/hour
Power source: electric
Price code: 4
AFOS Ltd, UK

BAKING OVENS
Travelling ovens for biscuits and bread making, and drying ovens.
Aifso Industrial Equipments Co., India

ROTARY OVEN
The temperature ranges from 50 to 300 degrees centigrade and the oven is programmable. It has 36 trays in two banks of 18, measuring 1.16 × 2 × 1.64 m. Power consumption is 3 kW/hour. Capacity: 1296 loaves/hour
Power source: electric
ANLIN, Peru

BAKERY MACHINERY
All types of bakery ovens.
Power source: diesel/gas/electric
Apple, India

ROTARY RACK OVEN
The circulation of hot air combined with the rotation of the trolley guarantees uniform and even baking. Constant temperature maintained through dual electronic thermostatic control. Capacity: 100–1000 loaves/hour
Power source: diesel/electric
Price code: 4
Baker Enterprises, India

ELECTRIC BAKING OVEN
Capacity: 850 kg/day
Power source: electric
Bijoy Engineers, India

BAKING OVEN
Two-deck electrically heated double-walled and fully insulated oven, complete with thermometric temperature control switch, pilot light and vapour escape.
Power source: electric
Price code: 3
Gardners Corporation, India

ROTARY BAKERY OVENS
Power source: gas
Industria Peruana Comercializadora Techno Pan Equipamiento Integral, Peru

ROTARY BAKERY OVENS
Power source: electric
Industria Peruana Comercializadora Techno Pan Equipamiento Integral, Peru

INJERA BAKING OVENS
Suitable for baking traditional *injera* and bread. Capacity: 30 pieces of *injera*/hour
Power source: electric
Price code: 1
Legio Aluminium and Electrical Bakery Industry, Ethiopia

ROTARY RACK OVEN
Uniform baking of all products. Heat retained during loading/unloading of trolleys. Can bake all varieties of bakery products such as bread, pastries, cakes, biscuits etc. Capacity: 168–700 loaves
Power source: electric
Prima Engineering Industries, India

OVENS
Industrial ovens, freeze dryers, multi-zone layered and conveyorized ovens, custom-built ovens and furnaces.
Shirsat Electronics, India

BAKERY OVENS
Power source: electric/gas/diesel
Technoheat Ovens and Furnaces Pvt Ltd, India

ENERGY-CONSERVING COOKER
Capacity: 5–10 kg/hour
Power source: biomass
Price code: 1
Usa Patanasetagit Company Ltd, Thailand

BAKING OVEN MODEL 710 8PX
138 × 90 × 83 cm. Complete with thermostat and safety devices.
Power source: gas
Well Done Metal Industries, The Philippines

SUPER BABY MINI OVEN
108 × 70 × 79 cm. Complete with thermostat, safety failure device and manual firing of burner.
Power source: gas
Well Done Metal Industries, The Philippines

ELECTRIC COOKING OVEN
Power source: electric
Price code: 3
Xuan Kien Private Company, Vietnam

25.1 Stoves

MULTIFUEL COOKING STOVE
Simple, portable cooking stove which uses biomass, such as wood, coal, crop residues, briquette, stalks etc., as fuel.
Power source: biomass
Central Institute of Agricultural Engineering, India

25.2 Integrated baking machines

CONTINUOUS CHAPATTI MAKING PLANT
Chapatti is an unleavened, baked and puffed staple food from India. This machine is a forming unit and baking unit. The forming unit converts wheat dough into disks of 120–150 mm diameter and 1 mm thick. The baking unit bakes both sides of the disks on a continuously moving carbon steel pan, heated from below by petroleum gas fired burners. Capacity: 600–800 chapattis/hour
Central Food Technological Research Institute, India

CONTINUOUS DOSA MAKING MACHINE
Dosa is a traditional Indian cereal-based pancake. The machine consists of a carbon steel rotating plate, heated from below by a petroleum gas burning system. There is a product scraping/rolling facility for detaching formed *dosas* from the hotplate. Capacity: 400 pieces per hour
Central Food Technological Research Institute, India

CONTINUOUS IDLI MAKING MACHINE
Idli is a traditional Indian breakfast food. This machine can be used for continuous cooking of any cereal-based food of specific shape. It has a stainless steel tunnel in which a conveyor with non-sticky pans holding *idli* batter travels. The tunnel has steam heating arrangements. There is also a product scooping facility. Capacity: 1200 pieces per hour
Price code: 4
Central Food Technological Research Institute, India

26.0 Packaging equipment

Packaging materials and various types of packaging machinery, including labelling machines, are available for a range of commodities.

CARTON PACKAGING EQUIPMENT
To hot-fill juice and to automatically form, fill and seal various sizes (0.25, 0.5 and 1 litre) of plastic-coated paper board gable-top cartons. Capacity: 2–100 cartons/minute
Power source: electric
Price code: 4
Alvan Blanch, UK

FILLING AND PACKING MACHINES
A range of fillers for bottles, cups and cartons; sealing and capping systems.
Charles Wait (Process Plant) Ltd, UK

TEA BAG PACKING MACHINE
1.8 hp/1700 rpm motor. Packs herbal teas and similar products. Capacity: 100–120 tea bags/minute
Power source: electric
Price code: 4
Industrias Technologicas Dinamicas SA, Peru

PACKAGING MACHINES
Packaging machines such as heat-sealers, carton-sealing machines, stretch-wrapping machines, case packers and palletizers.
Orbit Equipments Pvt Ltd, India

PACKAGING EQUIPMENT
Specialists in packaging automation for solid, powder and liquid.
Sunray Industries, India

PACKAGING EQUIPMENT
Packs candy, biscuits and other similar foods into pillow form. Size of machine: 3700 × 800 × 1500 mm. Vertical and horizontal heater. The equipment can be used with many kinds of materials as well as sizes of package: 50–200 mm length; 15–150 mm width; 2–30 mm height. It is made of stainless steel and rustproof material. Capacity: 100–250 packs/minute
Power source: electric
Price code: 3
Technology & Equipment Development Centre (LIDUTA), Vietnam

PACKAGING EQUIPMENT
Used for packing food in powder, grains/granules, liquid form. Vertical and horizontal heater. The equipment can be used with many kinds of materials as well as sizes of package: 150–350 mm length; 130–250 mm width. Capacity: 20–50 bags/minute
Power source: electric
Price code: 3
Technology & Equipment Development Centre (LIDUTA), Vietnam

26.1 Sealing equipment

AUTOMATIC CONVEYORIZED HOT AIR SACHET SEALER
Used for the perfect end sealing of pre-made pouches filled with edible products. The sealing height is adjustable allowing different pouches of pre-made sizes to be perfectly sealed with the aid of hot air free from contamination.
Price code: 3
Acufil Machines, India

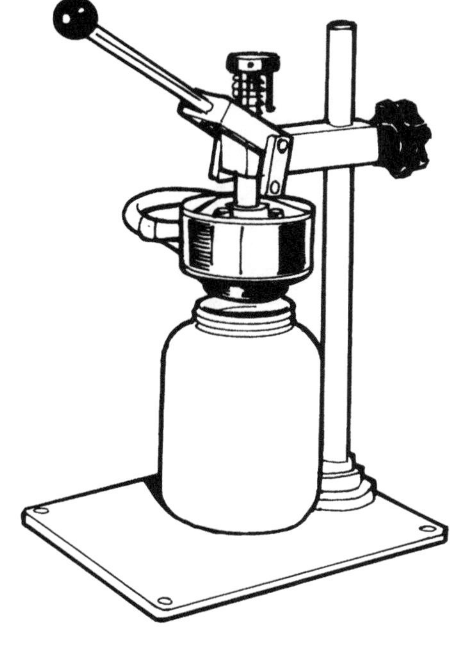

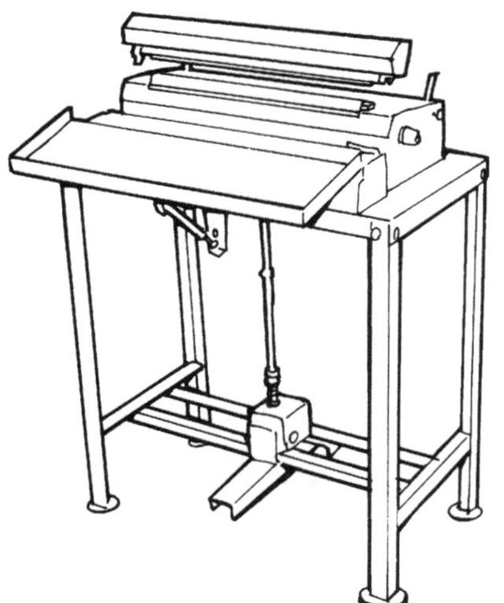

FORM FILL AND SEAL MACHINE
Has adjustable volumetric cups to allow for variation in bulk density between batches. Used for the sachet packing of all granular food products, such as dhal, peanuts and cashew nuts. Capacity: 20–40 sachets/minute
Power source: battery/electric
Price code: 4
Acufil Machines, India

BOTTLE WASHING, FILLING, CAPPING/SEALING MACHINE
This combination machine can operate automatically or semi-automatically and is suitable for a range of bottles from 0.5 to 1.5 litre volume. Washing operation is intermittent. Filling and capping/sealing operation is continuous. Low percentage of defective products. Capacity: 1000–2000 bottles/hour
Power source: electric
Price code: 4
ALFA Technology Transfer Centre, Vietnam

HEAT SEALERS
Used for the sealing of plastic bags.
Price code: 2/3/4
Alvan Blanch, UK

PLASTIC FILM HEAT SEALING MACHINE
Foot operated. Capacity: 120 bags/hour
Power source: electric
Price code: 1
Banyong Engineering, Thailand

BAG CLOSING MACHINE
Hook and bottom platform to hold the bags.
Power source: electric
Price code: 1
Gardners Corporation, India

ELECTRIC BREAD WRAPPER SEALING MACHINE
Complete with thermostatic switch.
Power source: electric
Price code: 1
Gardners Corporation, India

ELECTRIC POLYTHENE BAG SEALING MACHINE
Foot operated, and incorporates temperature as well as impulse control (to avoid burning of bags).
Power source: electric
Price code: 1
Gardners Corporation, India

FORM FILL SEAL MACHINE
A full automatic vertical form fill-seal machine for producing sachets/pouches from heat-sealable laminate. Used for packing any free-flowing powder, granules, spices, soup concentrates, soft drink concentrates, tea, coffee etc.
Gurdeep Packaging Machines, India

IMPULSE HEAT SEALER
The sealer is operated by foot, leaving both the hands free to handle the film. The bottom 'jaw' of the sealer containing the heated wire sealing strip is mounted on the work table. Depressing the foot pedal causes the rectangular steel frame to move downwards bringing the upper sealing head, mounted on the base of the moving frame, into contact with the bottom sealer. At this point a pin pushes against a simple push switch allowing current to flow and heat the sealing element. The work table is marked so that bags of different sizes can be made.
Power source: electric
Price code: 1
John Kojo Arthur, Ghana

MANUAL HEAT SEALING MACHINE
Used for sealing plugs of plastic containers, bottles, jars and jerrycans.
MMM Buxabhoy & Co., India

BOTTLE SEALING MACHINE
Capacity: 1000+/hour
Power source: manual/electric
Pharmaco Machines, India

Sealing equipment is also manufactured by Rieco Industries Ltd, India (Automatic Tube Filling and Sealing Machine), and Bajaj Maschinen Pvt Ltd, India (Bottle Sealer).

26.2 Capping equipment

CROWN CORKING MACHINE
This machine is adjustable for all sizes of bottles and has a press type cork holder.
Power source: manual
Price code: 1
Gardners Corporation, India

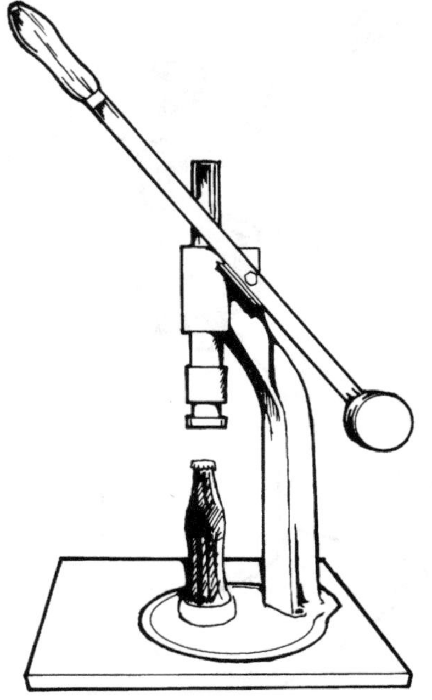

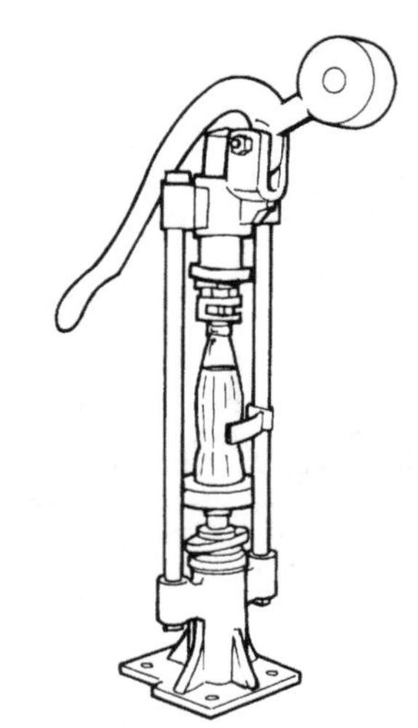

CROWN CORKING MACHINE
Foot operated with cork bowl and magnet crown holder.
Power source: manual
Price code: 1
Gardners Corporation, India

PILFER PROOF CAP SEALING MACHINE
Has an automatic sealing operation (one turn of the crank threads the cap and seals the end).
Price code: 1
Gardners Corporation, India

CAPPING MACHINE
Capacity: 3000–4000 per day
Power source: manual
Rank and Company, India

Capping equipment is also manufactured by Narangs Corporation, India.

26.3 Wrapping equipment

TEA-BAG OVERWRAPPER
Machine used for wrapping tea-bags.
Industrias Technologicas Dinamicas SA, Peru

Wrapping equipment is also manufactured by M/S Wonderpack Industries (P) Ltd, India (Shrink Wrapping Tunnels and Systems); Plasticam, Cameroon (Plastic Cartons); and Printpak, Cameroon (Cartons).

26.4 Vacuum packaging equipment

Vacuum Forming Machinery is manufactured by M/S Wonderpack Industries (P) Ltd, India.

26.5 Labelling machines

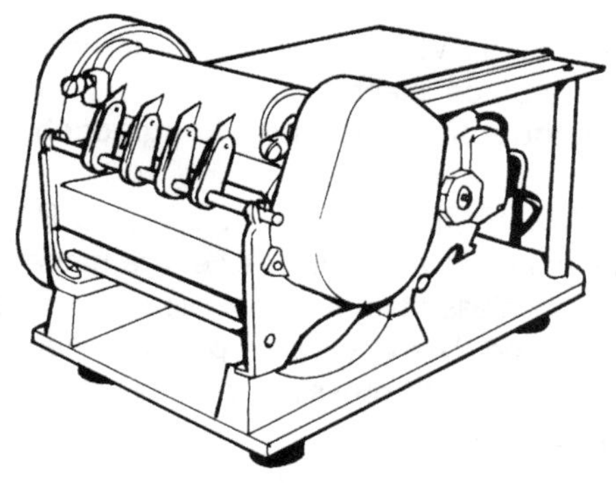

SEMI AUTOMATIC LABELLING MACHINE
Suitable for all types of round containers, jars, tins, cans and bottles. Capacity: 30–40 containers/minute
Bhavani Sales Corporation, India

LABEL GUMMING MACHINE
This hand-operated label gumming machine is suitable for labels of up to 15 cm width.
Power source: manual
Narangs Corporation, India

LABEL GUMMING MACHINE
Capacity: 1000+/hour
Pharmaco Machines, India

LABEL GUMMING MACHINE
Used for pasting gum on labels.
Rank and Company, India

26.6 Canning equipment

CAN CLOSING MACHINE
This hand-operated can sealer is flywheel operated and has automatic sealing operation capable of sealing cans up to 10 cm in diameter.
Power source: manual
Narangs Corporation, India

Canning equipment is also manufactured by Cantech Machines, India (Can Seaming and Reforming Machines).

26.7 Packaging materials

PACKAGING EQUIPMENT
Manufacturers of corrugated rolls, boards, boxes, cartons and containers etc.
Sridevi Packing Industries, India

Packaging materials are also manufactured by Ere du Verseau, Benin (Packaging); Fabasem, Cameroon (Box Cartons); Ghana Carton Boxes Mfg Ltd, Ghana; Helepac, Cameroon (Plastic Pots); Papier Plus, Cameroon (Paper Sachets and Cartons); Plastics Packaging Products Ltd, Ghana (Plastic Pots); Sada-SA, Mali (Plastic

Bottles); Top Industrial Packaging Products Ltd, Ghana (Plastic Containers); and Tropical Glass Co Ltd, Ghana (Glass Bottles).

27.0 Pans, kettles and boilers

A range of cooking pans, pots, vessels and kettles are available for various food groups and processing operations.

JAM BOILER
Suitable for the cooking of jam. Capacity: from 400 kg/day
Power source: electric
Alvan Blanch, UK

SUGAR BOILERS
Heavy duty stainless steel fabricated pans mounted on mild steel waste/wood fired furnace with tall chimney. High efficiency furnace and grate system. Complete with integral water tank and hose system for plant cleaning. Boils the juice and provides efficient separation during the flocculation process. Also condenses the juice into syrup.
Price code: 4
Alvan Blanch, UK

BABY BOILER
Boiler for small-scale use.
Azad Engineering Company, India

PAR BOILING PLANTS
Used for boiling raw paddy. Capacity: 250–2500 kg
Power source: electric/diesel
DIW Precision Engineering Works, India

KHOA MACHINE
Sweets that need heating and stirring can be prepared very easily using this machine. It can also be used for roasting various masala ingredients and for preparing various fruit juices. A detachable vessel allows the furnace to be used separately for other purposes. Optional ball valve can be supplied for removing liquid ingredient. Capacity: 15–60 litres
Power source: gas/diesel
Solar Arks, India

BABY BOILER
Boiler for small-scale use.
Geeta Food Engineering, India

STEAM COOKING VESSELS
A range of steamers that are suitable for various food groups.
Sri Rajalakshmi Commercial Kitchen Equipment, India

COOKING KETTLE ELECTRICAL
Complete with automatic digital temperature controller which can be set to desired temperature. Capacity: 125 kg/hour
Power source: electric
Price code: 2
Tinytech Plants, India

27.1 Steam jacketed pans and kettles

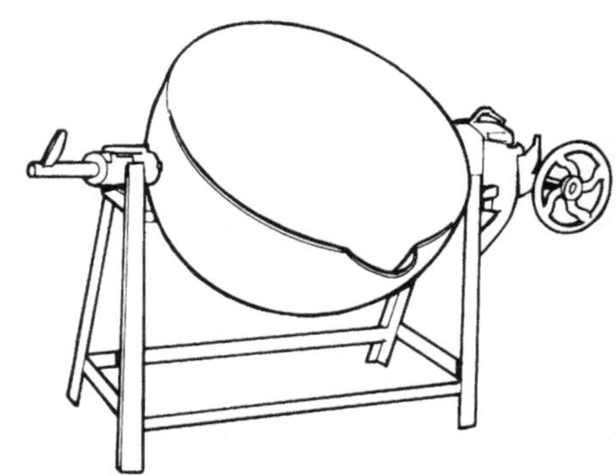

STEAM JACKETED KETTLE
Capacity: 227 litres
Price code: 2
Gardners Corporation, India

JACKETED VESSEL WITH STIRRER
Power source: electric
Mark Industries (Pvt) Ltd, Bangladesh

STEAM JACKETED TYPE TILTING PAN
The inner pan and outer jacket are made from stainless steel and mounted on a mild steel stand. There is also a pressure gauge, safety valve and pressure release valve.
Capacity: 10–50 kg
Narangs Corporation, India

BABY BOILER
Provides steam to cooking kettle to cook oilseeds.
Power source: biomass
Price code: 2
Tinytech Plants, India

STEAM COOKING KETTLE
Capacity: 125 kg/hour
Price code: 2
Tinytech Plants, India

Steam jacketed kettles are also manufactured by Bombay Industrial Engineers, India, and Techno Equipments, India.

27.2 Fryers

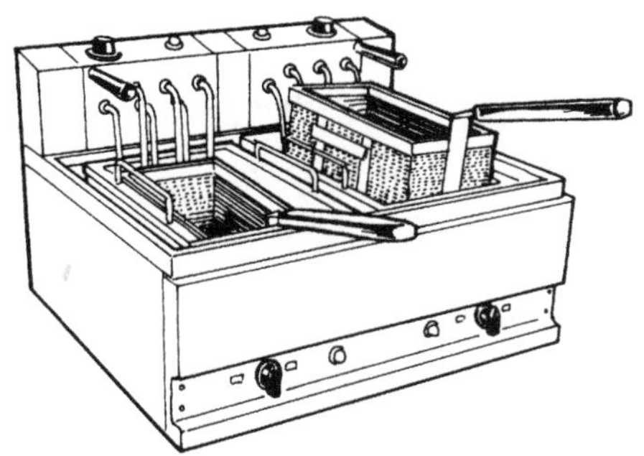

FRYER FOR SNACK FOODS – MULTI PURPOSE
Capacity: 100 kg/hour
Price code: 4
Crypto Peerless Ltd, UK

DEEP FAT FRYER
Temperature control thermostat.
Power source: electric
Price code: 2
Gardners Corporation, India

FRYER FOR SNACK FOODS – MULTI PURPOSE
Capacity: 100 kg/hour
Power source: electric/diesel
Price code: 4
Wintech Taparia Ltd, India

28.0 Pasteurizers

Pasteurization is a simple process to extend the shelf life of various commodities such as milk, juices and sauces. Heating the food to temperatures below 100 °C for different lengths of time, depending upon the food, destroys the micro-organisms present without causing significant changes to the food quality, texture and appearance. There are two types of pasteurization – bulk processing, in which the food is pasteurized in boiling kettles before being put into the containers, and in-container pasteurization, where the packaged product is pasteurized.

JUICE PASTEURIZERS
Power source: electric
Price code: 4
Alvan Blanch, UK

PASTEURISING UNITS
Designed for small-scale heat treatment of liquid products. The unit is used for the pasteurization of milk and cream, but can also be used for other products. The unit is provided with an electrical water heater and can be used where steam or hot water is not available. Capacity: 200–1000 litres/hour
Power source: manual
APV Unit Systems, Denmark

LABORATORY PASTEURIZER
Has a unique miniature plate heat exchanger with heating, regeneration and cooling sections. Capacity: 0–30 litres/hour
Power source: electric
Armfield Ltd, UK

MULTI-PURPOSE PROCESSING VESSEL
The facilities to mix, emulsify, heat, pasteurize, incubate, cool, chill and cure are all built in to this purpose-designed unit, a fully self-contained system in a mobile cabinet. Capacity: 10–20 litres
Power source: electric
Armfield Ltd, UK

PASTEURIZERS
A range of pasteurizers and heat exchangers for small- and large-scale food processing.
Charles Wait (Process Plant) Ltd, UK

PASTEURIZER – YOGHURT/CUSTARD KETTLE
Vat for making yoghurt. Hot and cold water can be circulated between the two inner walls of this triple walled vat, enabling the milk to be both heated and cooled. Capacity: 50–1000 litres
Power source: electric
C. van 't Riet Zuiveltechnologie BV, The Netherlands

VETERINARY/CATTLE & CAN WASHING EQUIPMENTS & UTENSILS
A range of canning equipment, including can steaming blocks, can washing troughs and mini-pasteurizers.
Dairy Udyog, India

PACKO BATCH PASTEURIZER
Specially developed for producing smaller quantities of milk, cream, yoghurt, cheese and cultured cream. Capacity: 150–500 litres/hour
Power source: electric
Fullwood Ltd, UK

MAYOTTE PASTEURIZER
Manual pasteurizer for juice, jelly, fruit and vegetables.
Power source: manual
Gilson Père et Fils, France

PASTEURIZERS
Capacity: 500–20 000 litres/hour
Power source: electric
Goma Engineering Pvt Ltd, India

PASTEURIZER
Power source: electric
Mark Industries (Pvt) Ltd, Bangladesh

MILK PASTEURIZER
Power source: electric
RPM Engineers (India) Ltd, India

Pasteurizers are also manufactured by Actini, France; Alven Foodpro Systems (P) Ltd, India; Brasholanda SA – Equipamentos Industriais, Brazil; Gilson Père et Fils, France; and Machin Fabrik, India

29.0 Peeling equipment

Manually operated and powered peeling equipment is available.

COCONUT SKIN PEELING MACHINE
Capacity: 50 kg/hour
Power source: electric
Price code: 1
Bay Hiap Thai Company Ltd, Thailand

POTATO PEELER
Used for removing the thin skin of potatoes and carrots etc.
B Sen Barry & Co., India

POTATO PEELER
Suitable for peeling any size potatoes for further processing into chips etc. Suitable for small-scale processing units. Capacity: 30–35 kg/hour
Power source: manual
Price code: 1
Central Institute of Agricultural Engineering, India

PINEAPPLE PEELER
Pedal-operated pineapple peeler used to peel and remove the core.
Power source: manual
Price code: 2
DISEG (Diseno Industrial y Servicios Generales), Peru

POTATO PEELER
Capacity: 100–150 kg/hour
Power source: electric
Price code: 2
Gardners Corporation, India

POTATO PEELER
1.8 hp/1700 rpm motor. Capacity: 350–400 kg/hour
Power source: electric
Price code: 3
Industrias Technologicas Dinamicas SA, Peru

VEGETABLE PEELER/WASHER
Peels the skin of vegetables and simultaneously washes. Capacity: 60–100 kg/hour
Power source: electric
Rajan Universal Exports (Manufacturers) Pvt Ltd, India

PEELER
Suitable for home or group use.
Robot Coupe, France

POTATO PEELER
Capacity: 180 kg/hour
Power source: electric
Servifabri SA, Peru

POTATO PEELER
Capacity: 700 kg/hour
Power source: electric
Price code: 3
Servifabri SA, Peru

ABRASIVE POTATO PEELER
Available in two sizes. Capacity: 180 and 700 kg/hour
Power source: electric
Price code: 3
Servifabri SA, Peru

Peeling equipment is also manufactured by Crypto Peerless Ltd, UK.

30.0 Presses

Presses are used for the extraction of liquids such as oil or fruit juice. A range of manually operated and powered fruit- and oil-pressing equipment is available.

SUGAR PRESS
Suitable for the crushing of sugar cane and for the extraction of the juice. Fitted with auto-scraper gear, oil lubrication systems, feed and discharge chutes, and double reduction gearing.
Power source: electric
Price code: 4
Alvan Blanch, UK

SCREW TYPE PANEER PRESSING DEVICE
Presses coagulated soya paneer to cubical form. Capacity: 16 kg/hour
Power source: manual
Central Institute of Agricultural Engineering, India

P2R 500 CASSAVA PULP PRESS
Used after grating and fermenting to mechanically extract water from the cassava pulp, to reduce the moisture level to a suitable rate for further processing. The village model is based on a hydraulic system (PH series).
Capacity: 0.1–3 tonnes/hour of fresh roots
Power source: electric
Gauthier, France

HORIZONTAL PLATE FILTER PRESS
Available in two models of 6 and 8 plates up to 20 inch diameter.
Rank and Company, India

30.1 Fruit and vegetable presses

Powered

JUICE PRESS
Press for extracting fruit juices.
Altech, France

FRUIT PROCESSING EQUIPMENT
Fruit processing equipment for making fruit and vegetable juice (300–5000 litres/hour); fruit concentrates (minimum 300 kg/hour); jams (at least 200 kg/hour) and vegetable pickles with vinegar.
Biaugeaud Henri, France

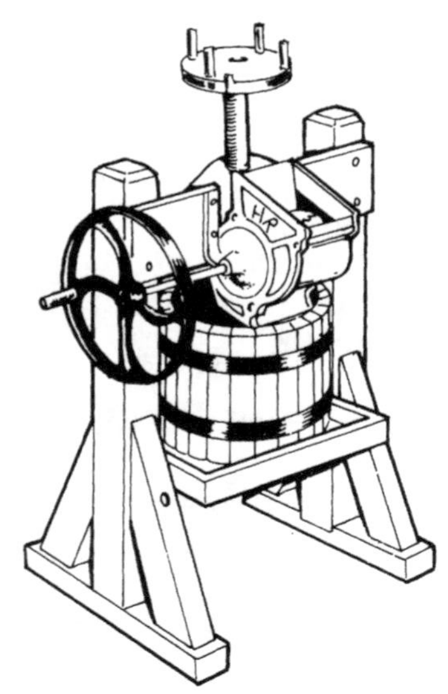

HYDRAULIC JUICE PRESS
Pressure ranges from 25 to 500 tonnes.
B Sen Barry & Co., India

FRUIT AND VEGETABLE MACHINERY
Fruit and vegetable preparatory, processing and canning equipment, fruit juice concentration equipment and filling machines.
Jwala Engineering Company, India

FRUIT PRESSES
Presses and crushers for apples, pineapples and mangoes.
Société Pontchatelaine Equipement Cidricole, France

PALM FRUIT POUNDING MACHINE
This machine pounds the fruit in a continuous operation. Capacity: 25 kg/minute
Power source: diesel/electric
SSIS Enterprises, Ghana

Manual

SELF CONTAINED FRUIT PRESSES
Has a built-in self-feed grinder with an extra heavy press mechanism and oversize basket. Mashes all fruits with cast iron flywheel assisted crank.
Power source: manual
Price code: 2
Lehman Hardware & Appliances, USA

HAND SCREW BASKET PRESS
Has a steel frame, wooden tray and basket, and extracts juice from soft fruits.
Price code: 2
Narangs Corporation, India

TOMATO AND GRAPE CRUSHER
This machine will crush tomatoes and other soft fruits. The hopper is suitable for hand or pulley drive.
Price code: 3
Narangs Corporation, India

Fruit and vegetable presses are also manufactured by Bajaj Maschinen Pvt Ltd, India, and Bombay Industrial Engineers, India.

30.2 Oil presses

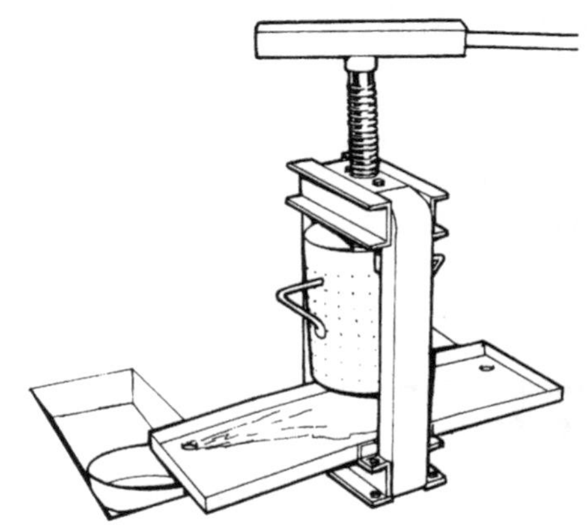

Powered

COMBINATION OIL SCREW PRESS
Two to ten units of the Mini 50 Oil Screw Press can be arranged to operate together to give higher outputs. Presses are fed by a mechanical conveying system with in-built preheating. Capacity: 500 kg/hour
Power source: electric
Alvan Blanch, UK

FILTER PRESS
Used to remove the majority of solid particles from crude oil to obtain clarified oil.
Alvan Blanch, UK

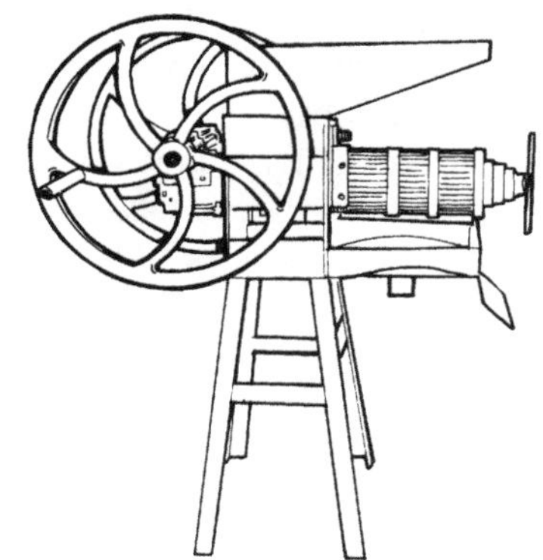

KOMET DOUBLE SPINDLE PRESS
Used for the extraction of vegetable oil by cold-pressing from oil-seeds, kernels and nuts. Capacity: 25–70 kg/hour
Power source: diesel/electric
IBG Monforts GmbH & Co., Germany

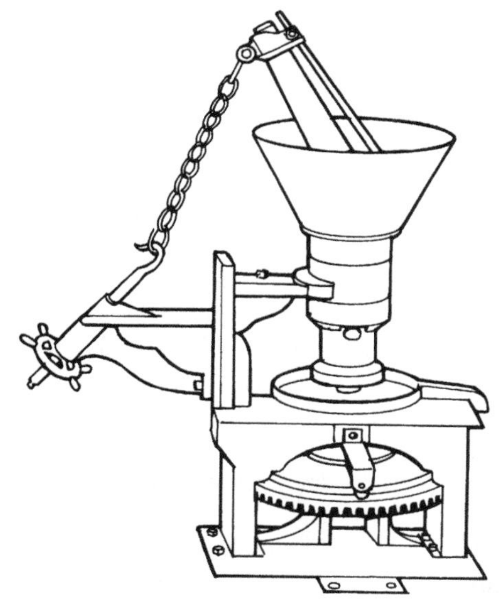

KOMET SINGLE SPINDLE PRESS
Designed for use in the laboratory/research sections of edible oil and margarine factories.
Capacity: 8–15 kg/hour
Power source: electric
IBG Monforts GmbH & Co., Germany

OIL PRESS
Part of a range of husking machines, presses, filter presses, crushers and complete units from traditional to industrial oil mills.
La Mécanique Moderne, France

OIL PRESS
Power source: manual
Rajan Universal Exports (Manufacturers) Pvt Ltd, India

COCONUT MILK EXTRACTOR
Capacity: 35 kg/hour
Power source: electric
Price code: 3
Saw Charoenchai, Thailand

PALM OIL PRESS
Used to press/squeeze palm oil from the fibre or to extract the oil from the pounded fruit.
Technology Consultancy Centre, Ghana

SHEA PASTE KNEADING MACHINE
Capacity: 1000 kg/hour
Power source: diesel/electric
WA ITTU, Ghana

Manual

RAM PRESS
A manually operated semi-continuous oil-seed press suitable for rural processors.
Power source: manual
Agro Alfa, Mozambique

MANUAL OIL PRESS
Capacity: 9 kg/hour
Power source: manual
Alvan Blanch, UK

RAM PRESS
A manually operated semi-continuous oil-seed press suitable for rural processors.
Power source: manual
ApproTEC, Kenya

RAM PRESS
This press is a manually operated semi-continuous oil-seed press suitable for rural processors.
Power source: manual
Centre for Agricultural Mechanisation and Rural Technology (CAMARTEC), Tanzania

BRIDGE PRESS
This machine is used for extracting oil from coconuts. Height of cage 40 cm; diameter 24.2 cm. Capacity: 200 coconuts in 8 hours
Power source: manual
Price code: 2
Coconut Development Authority, Sri Lanka

OIL PRESS FOR PEANUTS
A manual oil press for peanuts which is made of painted steel with 4 inch girder supports, 2 inch square threaded spindle and positioning feet. It has a perforated steel cage and an oil collecting receiver.
Power source: manual
Price code: 2
DISEG (Diseno Industrial y Servicios Generales), Peru

RAM PRESS
A manually operated semi-continuous oil-seed press suitable for rural processors.
Power source: manual
Karume Technical College, Tanzania

RAM PRESS
A manually operated semi-continuous oil-seed press which can be easily operated by individuals to extract oil from copra and coconut. Capacity: 4 kg/hour
Power source: manual
Natural Resources Institute, UK

SHEA NUT JACK PRESS
Used to press shea nuts to produce vegetable oil. Portable. Production capacity reaches its maximum if a grinder is used initially to reduce the seeds to paste or small particles. Capacity: 40 litres/hour
Power source: manual
Northern Uganda Manufacturers Association, Uganda

BRIDGE PRESS
Used for extracting oil from coconuts. Height of cage 40 cm; diameter 24.2 cm. Capacity: 200 coconuts in 8 hours
Power source: manual
Price code: 2
Société Ivoirienne de Technologie Tropicale, Côte d'Ivoire

BRIDGE PRESS
Used for extracting oil from coconuts. Height of cage 40 cm; diameter 24.2 cm. Capacity: 200 coconuts in 8 hours
Power source: manual
Price code: 1
Space Engineering Ltd, Tanzania.

BRIDGE PRESS
Used for extracting oil from coconuts. Height of cage 40 cm; diameter 24.2 cm. Capacity: 200 coconuts in 8 hours
Power source: manual
Price code: 2
Technology Consultancy Centre, Ghana

Oil Presses are also manufactured by Altech, France, and Outils Pour Les Communautes, Cameroon.

31.0 Pulpers and juicers

Pulpers and juicers are used for the extraction of fruit and vegetable juices for the preparation of jams, jellies, sauces, fruit juices and cordials. Coffee pulpers, which are used during processing of coffee beans, are also available.

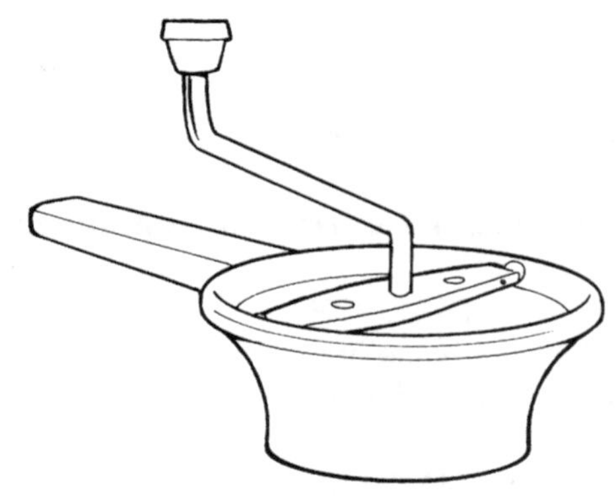

ORANGE JUICE EXTRACTOR
Efficiently extracts the juice from oranges without contamination from the peel.
Capacity: 20 oranges/minute
Price code: 4
Alvan Blanch, UK

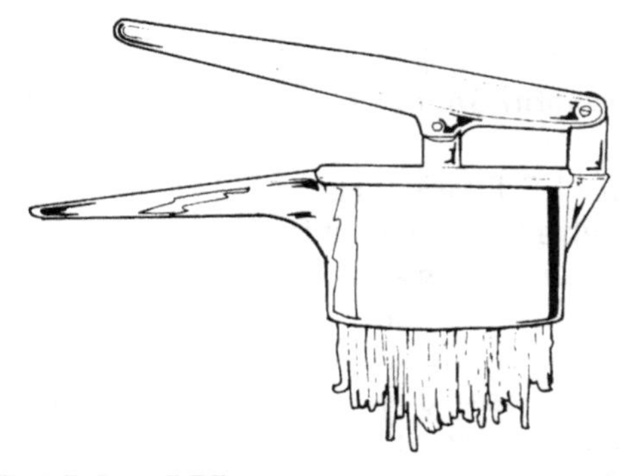

FRUIT PULPER
Can be used as a pulper and as a sieve machine. It has two mesh sizes – 3 mm and 0.5 mm. Capacity: 40–50 kg/hour
Power source: electric
Price code: 3
DISEG, Peru

PULPER
Capacity: 300 kg/hour
Power source: diesel
DISEG, Peru

PULPER
Capacity: 100–2000 kg/hour
Eastend Engineering Company, India

SCREW TYPE JUICE EXTRACTOR
Capacity: 25–100 kg/hour
Eastend Engineering Company, India

CONE TYPE JUICER
Juicer for tight skin oranges.
Power source: electric
Price code: 1
Gardners Corporation, India

ELECTRIC MULTIPRESS JUICER
Suitable to juice all fruits and vegetables, including apples, pears, carrots and pineapples.
Power source: electric
Price code: 1
Gardners Corporation, India

JUICER
Hand wheel juicer – cone type for oranges and lemons.
Price code: 1
Gardners Corporation, India

PULPER
Capacity: 0.5 tonne/hour
Power source: electric
Price code: 2
Gardners Corporation, India

PULPER JUNIOR MODEL
Complete with sieve, covers, shaft and hopper. Capacity: 0.5–1 tonne/hour
Power source: electric
Price code: 2
Gardners Corporation, India

SCREW JUICER
Power source: electric
Price code: 1
Gardners Corporation, India

TOMATO AND GRAPE CRUSHER
Has a hopper mounted on a stand.
Price code: 1
Gardners Corporation, India

FRUIT JUICE EXTRACTION MACHINE
Capacity: 150 kg/hour
Power source: electric
Price code: 2
Kasetsart University, Thailand

PULPER
Power source: electric
Price code: 2
Kundasala Engineers, Sri Lanka

PULPING MACHINE
Capacity: 30–35 kg/hour
Power source: electric
Mark Industries (Pvt) Ltd, Bangladesh

COIL TYPE JUICE EXTRACTOR
Has an aluminium juicing head with a stainless steel sieve. Capacity: 750–2000 oranges/hour
Power source: electric
Price code: 2
Narangs Corporation, India

FRUIT AND VEGETABLE PULPER
Capacity: 80 kg/hour
Power source: electric
Narangs Corporation, India

CITROZYM L, 100 L AND CITROZYM G
Suitable for all steps of citrus processing – peel extraction, pulp wash, viscosity reduction, juice clarification, oil recovery and fruit peeling.
Novo Nordisk Ferment Ltd, Switzerland

JUICE EXTRACTOR
Suitable for home or group use.
Robot Coupe, France

PULPING MACHINE
Capacity: 95 kg/hour
Power source: electric
Price code: 3
Sahathai Sathorn, Thailand

FRUIT PULPER/SIEVE MACHINES
Capacity: 10–500 kg/hour
Power source: electric/manual
Price code: 2/3
Servifabri SA, Peru

PULPER FINISHER
Suitable for a range of fruits. It is available in a range of sizes and can be manually operated or power driven. The powered versions are available in 1 hp, 2 hp (both single phase), 3 hp and 6 hp (both triple phase). Capacity: 10, 20, 100, 200, 500 kg/hour
Power source: manual/electric
Price code: 2/3
Servifabri SA, Peru

FRUIT PULPERS
1–5 kW engine. Capacity: 100–200 kg/hour
Power source: electric
Udaya Industries, Sri Lanka

Pulpers and juicers are also manufactured by: Bajaj Maschinen Pvt Ltd, India (Fruit Mill Crusher, Pulper); Bombay Industrial Engineers, India (Pulpers); Geeta Food Engineering, India (Pulper-cum-Finisher, Juice Extractors, Fruit and Vegetable Crusher); Narangs Corporation, India

(Mincing Juicers/Pulpers); Sri Rajalakshmi Commercial Kitchen Equipment, India (Juice Machines); Techno Equipments, India (Fruit Pulpers).

31.1 Sugar cane crushers

CANE SUGAR JUICE EXTRACTION MACHINE
Used for extracting cane juice. Capacity: 50 litres/hour
Power source: electric
Price code: 2
Bay Hiap Thai Company Ltd, Thailand

SUGAR CANE CRUSHER
Used for extracting juice from sugar cane. Capacity: 75–750 kg/hour
Power source: electric
Rajan Universal Exports (Manufacturers) Pvt Ltd, India

SUGAR CANE CRUSHER
Suitable for the production of sugar cane juice. Capacity: 10 litres/hour
Power source: diesel/electric/manual
TEMDO (Tanzania Engineering and Manufacturing Design Organisation), Tanzania

31.2 Coffee pulpers

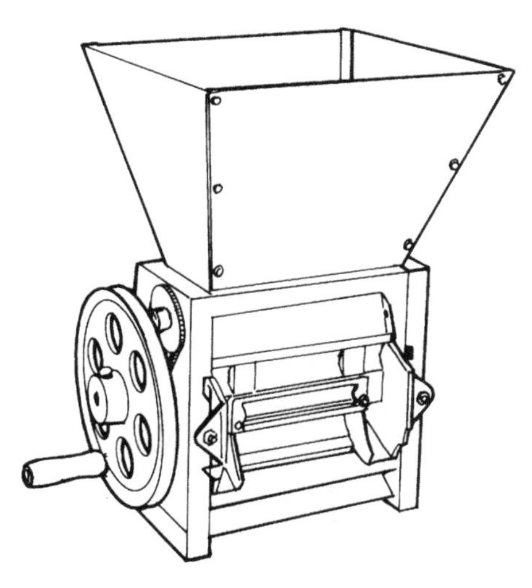

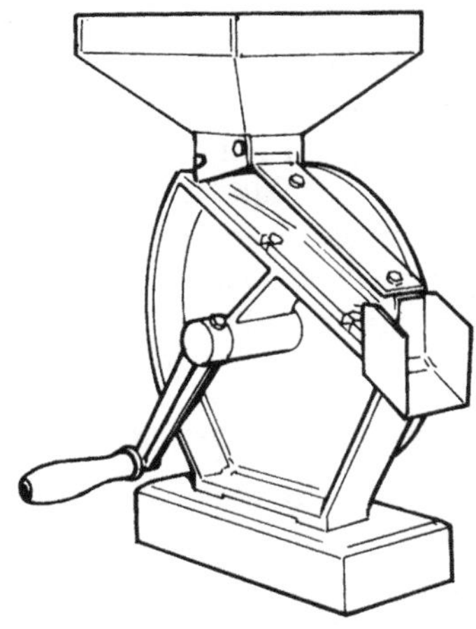

COFFEE PULPER
Has a large-capacity holding hopper and self-lubricating Oilite bushes. The pulping breast is easily adjustable to accommodate all types of coffee. Capacity: 300–4000 kg/hour
Power source: manual/electric/diesel/petrol
Price code: 2/4
Alvan Blanch, UK

COFFEE PULPERS
Used for depulping coffee cherries. The principal features of the drum pulpers are the large-capacity holding hopper and self-lubricating Oilite bushes. The pulping breast is easily adjustable to accommodate all types of coffee. Capacity: 300 kg/hour
Power source: diesel/electric/manual/petrol
Price code: 2/3
Denlab International (UK) Ltd, UK

'IRIMA 67' DISC PULPER
Used for depulping coffee cherries. Capacity: 270–360 kg/hour
Power source: manual
Price code: 2
John Gordon International, UK

COFFEE PULPER – DRUM PULPER
Used for pulping coffee beans on a small scale. Capacity: 300 kg/hour
Power source: manual/electric
Rajan Universal Exports (Manufacturers) Pvt Ltd, India

COFFEE PULPER
Removes skin from ripe coffee beans and can also be used for removing skin from neem and groundnuts. It has adjustable discs for different varieties and is fitted with ball bearings, flywheel, cover and handle etc.
Capacity: 40 kg/hour – manual
Udaya Industries, India

32.0 Roasting equipment

In its simplest form, foods are roasted on an open pan over any heat source. However, specialized roasting equipment, especially for coffee roasting, where the time and temperature of roasting need to be carefully controlled, is also available.

AUTOMATIC ROASTER – PAN TYPE
Used for the roasting of dhal, spices, wheat and all food products that require pre-cooking or pre-roasting.
Power source: gas/battery/electric
Price code: 3
Acufil Machines, India

AUTOMATIC ROASTER WITH ELECTRICAL HEATING PAD
Used for removing moisture or the roasting and heat impregnating processes. This system is totally covered and very hygienic.
Power source: electric/battery
Price code: 3
Acufil Machines, India

DRIED CHILLI ROASTING MACHINE
Capacity: 50 kg/hour
Power source: gas
Price code: 2
Bay Hiap Thai Company Ltd, Thailand

POULTRY ROASTING MACHINE
Power source: gas
Price code: 2
Bay Hiap Thai Company Ltd, Thailand

SEED ROASTING MACHINE
Capacity: 50 kg/hour
Power source: gas/electric
Price code: 2
Charoenchai Company Ltd, Thailand

GRAIN TOASTER WITH COOLING
The roasting cylinder is driven at 60 rpm by a 2.4 hp motor. The toasted product is cooled by air in a cone driven by a 1.8 hp motor.
Capacity: 100 kg/hour
Power source: electric
Industrias Technologicas Dinamicas SA, Peru

MULTI-PURPOSE SAND ROASTER
Traditionally, grain is roasted in very hot sand. This machine consists of a screen which only sand can pass through. By raising or lowering the screen, the grain is brought into contact with the hot sand for a predetermined time or removed from the sand. Capacity: 3 kg/hour
Power source: electric
Department of Agricultural Engineering, NERIST, India

Roasting equipment is also manufactured by Sri Murugan Industries, India, and Sri Venkateswara Industries, India.

32.1 Coffee Roasters

SEMI-AUTOMATIC COFFEE ROASTING MACHINE
Roasting drum made of stainless steel, 1000 mm in diameter, with a gas burner inside a heating cabinet. The gas burner provides heat to the roasting drum; gas flow is regulated through a valve which is controlled by an automatic temperature regulator. The motor and electrical system control the roasting process: speed, time, temperature.
Capacity: 70–100 kg/hour
Power source: electric
Price code: 3
ALFA Technology Transfer Centre, Vietnam

COFFEE ROASTER
A non-perforated roasting drum with a special paddle mixer assures an optimal blending and excellent coffee-roasting results. The roasting process can be easily controlled at all times through a sight glass, sampler tube and thermometer. A large cooling sieve with fan and stirring mechanism rapidly cools the coffee to the requested temperature. Capacity: 100 kg green coffee/hour
Power source: gas/electric
Price code: 4
Alvan Blanch, UK

ROASTING–COOLING MACHINE – TR-RF 250
Used for the discontinuous roasting of nuts or beans. Capacity: 15–500 kg/cycle
Gauthier, France

SMALL SCALE COFFEE ROASTER
Used for roasting batches of coffee beans but can also be used for roasting cashew nuts, peanuts, almonds etc. Comes with independent timer.
Power source: electric
Price code: 2
John Gordon International, UK

COFFEE MAKING EQUIPMENT
Roasters, grinders and mixers for coffee making.
Sri Venkateswara Industries, India

33.0 Smoking equipment

A range of equipment, including salting, cook stoves, smokers and drying rooms, is available for a range of foodstuffs, including meat, fish and cheese. Smokers are available in a range of styles and sizes to cater to all needs.

DOUBLE MAXI
Cooking, smoking and drying system. Designed for small batch processing of meat, fish and cheese etc. Three smoke box drawers are fitted at the left-hand side of the unit in which sawdust is placed and manually ignited. Capacity: 127 kg/hour
Power source: electric
Price code: 4
AFOS Ltd, UK

MAXI
Cooking, smoking and drying system. Designed for small batch processing of meat, fish and cheese etc. Three smoke box drawers are fitted at the left-hand side of the unit in which sawdust is placed and manually ignited. Capacity: 50 kg/hour
Power source: electric
Price code: 4
AFOS Ltd, UK

MINI
Smoking, cooking and drying system. Designed for small batch processing of meat, fish and cheese etc. Three smoke box drawers are fitted at the left-hand side of the unit in which sawdust is placed and manually ignited. Capacity: 25 kg/hour
Power source: electric
Price code: 4
AFOS Ltd, UK

EQUIPMENT FOR MEAT AND COOKED MEAT PROCESSING
Includes salting, cook stove, smoking and drying room equipment.
Capic, France

STOVE TOP SMOKER
This small smoker fits on top of a gas or electric hotplate. It can be used indoors to give a unique smoked flavour to fish, meat, poultry or cheese. It is made of heavy gauge stainless steel (45 × 28 × 9 cm) and weighs about 4 kg. Bulky foods can be cooked in the smoker. Smoke flavour is produced from 100 per cent natural virgin hardwood chips (included with the smoker).
Power source: electric/gas
Price code: 1
Sausage Maker Inc., USA

33.1 Smokehouses

STAINLESS STEEL 20 LB CAPACITY SMOKEHOUSE
This smokehouse is made of 24 gauge stainless steel and is insulated with 1 inch board so that temperatures of 160–170°F can be maintained. The smokehouse is equipped with three hickory wood dowels, four shelves, baby dial thermometer and stainless steel sawdust pan. The internal dimensions are 15"d × 20"h × 16.5"w. The power supply is a 120 V–1000 W heating element with adjustable control. Capacity: 10 kg per batch
Power source: electric
Price code: 2
Sausage Maker Inc., USA

20 LB INSULATED SMOKEHOUSE
This 20 lb capacity smoker comes complete with four wire shelves, three hickory hardwood dowels for hanging sausage, a baby dial thermometer and a stainless steel sawdust pan. The inside dimensions are 15"d x 20"h and 16.5" wide. It has a 6", 1000 W heating element that requires a current of 120 V. Capacity: 10 kg per batch
Power source: electric
Price code: 2
Sausage Maker Inc., USA

50 LB STAINLESS STEEL INSULATED SMOKEHOUSE
This smokehouse features a thermostat control with a +/-5 variation and also a 2½" dial thermometer that indicates the internal temperature of the smokehouse at all times. The cabinet is made of stainless steel (inside dimensions 21"w, 18"d, 32"h) and is insulated with heavy duty board. It is heated by a 2100 W/240 V heating element. The smoker is fitted with six stainless steel V-shaped sticks, a stainless steel pan (12 inch diameter) and three chrome-plated shelves. An insulated version of the 50 lb cabinet is also available. Capacity: 25 kg per batch
Power source: electric
Price code: 3
Sausage Maker Inc., USA

100 LB CAPACITY STAINLESS STEEL SMOKEHOUSE
A range of smokehouses is available – insulated and non-insulated made of different materials (aluminium, stainless steel). They are available with either gas or electric heaters. Capacity: 50 kg per batch
Power source: electric/gas
Price code: 4
Sausage Maker Inc., USA

34.0 Sorting and grading equipment

In most very small-scale operations, sorting of raw material and finished products is done by hand. However, for the larger scale there are a variety of small sorting machines available to grade foods on the basis of size, density or shape. Colour sorting machines are only available for large-scale operations and are expensive.

MECHANIZED GRADER
Used for separating food groups such as peanuts and dhal etc. into different sizes where grading by size is required. Capacity: 0.5–3 tons/hour
Power source: electric
Price code: 3
Acufil Machines, India

CLEAN-O-GRADER
Control of feeds facilitates finer grading of products of several varieties. Control of air allows suction of lighter impurities/seeds as desired. Pre- and post-aspiration system with collection chamber and cyclone ensures dust-free working in the plant. Capacity: 100–2000 kg/hour
Power source: battery/electric
Price code: 3
Goldin (India) Equipment (Pvt) Ltd, India

FRUIT SIZE GRADING MACHINE
Capacity: 270 kg/hour
Power source: electric
Price code: 3
Kasetsart University, Thailand

SCREENER
Screening machine and density separator for grains.
Marot, France

SIZING EQUIPMENT
Sizing equipment for coffee and cocoa.
Marot, France

POTATO GRADER MACHINE
Dimensions 5900 × 1750 × 2050 mm approximately. Capacity: 3500 kg/hour
Power source: electric
Mirpur Agricultural Workshop Training School (MAWTS), Bangladesh

GYRO SCREEN
Used for optimum continuous screening at maximum efficiency and feed rate. Feed material constantly flows in horizontal motion along a loop pattern on a vibro-screen surface. Capacity: 125–500 kg/hour
Power source: electric/diesel
Price code: 3
Premium Engineers Pvt Ltd, India

900TD WOODEN FRAME TESTING SCREENS
Used for the size classification of green coffee beans. The screens are manufactured from stainless steel and available with round hole perforations. The screens are supplied with nesting wooden frames, and optional top lid and bottom tray are available.
Power source: manual
Price code: 1
John Gordon International, UK

34.1 Separators

GRAIN–FLOUR SEPARATOR
Power-operated equipment for separation of milled wheat into various sizes which are collected at different outlets. Suitable for small-scale processing units. Capacity: 80–120 kg/hour
Power source: electric
Central Institute of Agricultural Engineering, India

SEPARATORS
A range of separators for small- and large-scale food processing.
Charles Wait (Process Plant) Ltd, UK

GRAVITY SEPARATOR
This machine grades the seeds on a gravity basis into heavies, mediums and lights so that the best quality seeds can be made available for export. Capacity: 500 kg/hour
Power source: diesel/electric
Price code: 3
Forsberg Agritech (India) Pvt Ltd, India

PRECISION AIR CLASSIFIER
Used for three functions – to separate hulls mixed with unhulled seeds; to recover oil bearing kernels from the hulls; and to separate light materials from heavy materials. Capacity: 150–2000 kg/hour
Power source: electric/diesel
Price code: 2/3
Forsberg Agritech (India) Pvt Ltd, India

TSR 150 SORTER
Can be used at different stages in rice processing: to separate long and short paddy grains; to eliminate large (lumps of earth etc.) or small (sand, seeds etc.) foreign matter; to remove white rice fragments; or to separate broken and whole grains. Capacity: 150–2000 kg/hour
Gauthier, France

GRAVITY SEPARATOR
Pressure/vacuum gravity separators.
Capacity: 100–2000 kg/hour
Power source: battery/electric
Price code: 3
Goldin (India) Equipment (Pvt) Ltd, India

SCREEN SEPARATORS
Circular motion screener is widely used for removal of oversized and undersized material. Capacity: 100–2000 kg/hour
Power source: battery/electric
Price code: 3
Goldin (India) Equipment (Pvt) Ltd, India

HI-CAP GRAVITY SEPARATOR
For separating dry bulk particles that are similar in size and shape but different in weight.
John Fowler (India) Ltd, India

PADDY SEPARATOR
Capacity: 750 kg/hour
Power source: electric
Price code: 2
Kundasala Engineers, Sri Lanka

VIBRO-SEPARATORS
Products include cleaners, graders, de-stoners, screens and separators.
Premium Engineers Pvt Ltd, India

34.2 Dairy separators

Cream separators are manufactured by Dairy Udyog, India, and Goma Engineering Pvt Ltd, India.

35.0 Soya processing equipment

This section includes equipment that is used specifically for the small-scale processing of soybeans into soya milk, tofu and soya paneer.

AGROLACTOR
Compact and automated platform for soya milk production. Capacity: 250 litres/hour
Power source: electric
Actini, France

COTTAGE SCALE SOYPANEER PLANT
A cottage-level soya paneer plant with machines including steam generation unit, grinder-cum-cooker, milk filtration unit and paneer pressing device. Capacity: 50 kg/day
Power source: manual
Central Institute of Agricultural Engineering, India

SCREW TYPE PANEER PRESSING DEVICE
Presses coagulated soya paneer to cubical form. Capacity: 16 kg/hour
Power source: manual
Central Institute of Agricultural Engineering, India

SOYBEAN DEHULLER
A unit specially designed to meet soybean de-husking and splitting requirement. Suitable for cottage to small-scale soybean processing units engaged in producing soya flour, milk, paneer and other soya products.
Capacity: 100 kg/hour
Power source: electric/manual
Central Institute of Agricultural Engineering, India

36.0 Storage equipment

A range of storage equipment, from plastic buckets and bins to metal tanks and vats, is available.

SACK HOLDER
Adjustable stand sack-holder which keeps sack in a vertical open position for easy loading of grains and other materials. The height can be adjusted depending upon the size of the sack. Capacity: 100 kg
Central Institute of Agricultural Engineering, India

INSULATED STORAGE TANK
Capacity: 500–20 000 litres
Goma Engineering Pvt Ltd, India

MILK PASTEURISATION PLANTS
Storage tanks for milk and equipment for making by-products from milk, such as khova, paneer, lassi and yoghurt. Ice cream equipment, vats and storage tanks for brewery, confectionery and cosmetic equipment.
Power source: electric
RPM Engineers (India) Ltd, India

Storage equipment is also manufactured by Charles Wait (Process Plant) Ltd, UK; RPM Engineers (India) Ltd, India; and Techno Equipments, India

37.0 Stuffing equipment

Stuffing equipment is mainly for the production of sausages. A full range of equipment and accessories, in a wide range of sizes, is available to cater to most needs.

SAUSAGE STUFFER ACCESSORIES
A range of accessories for use with the sausage stuffing machines: includes handles, cleaning brushes, 'O' rings, lock-nuts, pressure relief valves and jerky attachment.
Power source: manual
Price code: 1
Sausage Maker Inc., USA

DELUXE 25 LBS CAPACITY SAUSAGE STUFFER
This sausage stuffing machine has precision-made steel two-speed gears and self-lubricating nylon bushings. The cylinder and base are made of stainless steel. It is supplied with three different plastic stuffing tubes. Similar stuffers are available in different capacities – 15 lb (7.5 kg), 5 lb (2.5 kg). Capacity: 2.5 to 12.5 kg per batch
Power source: manual
Price code: 2
Sausage Maker Inc., USA

STAINLESS STEEL 1.25 L SAUSAGE STUFFER
This machine is used to make small amounts of sausage. It has three food-grade plastic stuffing tubes and is easy to clean with hot water and detergent. Capacity: 1.25 litres per batch
Power source: manual
Price code: 1
Sausage Maker Inc., USA

STAINLESS STEEL STUFFING TUBES
A range of stuffing tubes (plastic and stainless steel) available for use with the 15 lb (7.5 kg) and 25 lb (12.5 kg) sausage stuffers.
Price code: 1
Sausage Maker Inc., USA

3 LBS CAPACITY SAUSAGE STUFFER
Cast iron double tinned to prevent rust. Equipped with three plastic stuffing tubes – 5/8" for breakfast sausage, 3/4" for Polish sausage, 1" for large size salami. Also includes food-grade sleeve to prevent seepage of ground meat out of the side of the piston when stuffing meat. Capacity: 3 lb per batch
Manual
Price code: 1
The Sausage Maker Inc., USA

38.0 Testing equipment

Testing equipment is essential for quality control and assurance of certain foods. However, it is also important that the equipment is used by a trained operator.

MILK TESTING EQUIPMENT AND UTENSILS
Includes hand-operated and electric centrifuge machines, funnels, beakers, flasks and milk collecting trays, cans and pails.
Dairy Udyog, India

DENSITY METERS
Used for measuring the density of liquids.
Karishma Instruments Pvt Ltd, India

DATA LOGGERS, PRESSURE DEVICES, CONTROLLERS, INSTRUMENTS AND TRANSMITTERS
Full range of testing and measuring equipment.
Status Instruments Ltd, UK

ECM SYSTEM
Miniature manufacturing equipment for accurate tea research and development, quality assessment etc. Standard equipment includes Wither Cabinet, Ferment Cabinet, Fluid Bed Dryer, and Stalk Remover.
Teacraft Ltd, UK

OP-TEA-MIZER
Used for the accurate timing of black tea fermentation – to ensure that tea is processed at its peak value.
Teacraft Ltd, UK

38.1 Flow Meters

Flow meters are manufactured by Danfoss Ltd, UK.

38.2 Hydrometers

Hydrometers are used to measure the specific gravity of a liquid.

BRIX HYDROMETER
Price code: 1
Gardners Corporation, India

38.3 Moisture analysers

MOISTURE METERS
For measuring the moisture content of grains and pulses etc.
Power source: battery
Alvan Blanch, UK

THERMORAY
Used for simple and rugged moisture measurement of black tea from the dryer.
Teacraft Ltd, UK

38.4 Refractometers

Refractometers are used to measure the solids content (percentage of sugars) of products such as jams, jellies and juices. Specific honey refractometers are available for determining the fructose level of honey.

FISHERBRAND REFRACTOMETERS
Handheld, easy to use and accurate, with a focusing eye piece. Nine models are available, covering a range of applications and measurement scales. Some models have automatic temperature compensation (ATC) between 10 and 30 °C. The Brix scale is used to measure the sugar concentration in solutions such as fruit juice, wine and soft drinks. Different sizes are available as follows: 0–18 Brix; 0–32 Brix; 28–62 Brix; 45–82 Brix; 58–90 Brix.
Price code: 1
Fisher Scientific UK Ltd, UK

POCKET REFRACTOMETER
Comes with leather carrying case.
Price code: 1
Gardners Corporation, India

SUGAR REFRACTOMETER
Measures sugar content. The following size ranges are available: 0–45 per cent and 40–85 per cent.
Price code: 1
Narangs Corporation, India

Barometers are manufactured by West Meters Ltd, UK.

38.5 Temperature measurement

There are three basic types of thermometer – those that contain mercury or alcohol in a glass tube, panels that change colour with different temperatures, and electronic or digital thermometers. The first are cheaper and do not require maintenance but are easily broken, which is hazardous in a food processing area. Mercury thermometers should never be used in a food processing room. Electronic thermometers are more expensive and difficult to maintain, but are able to measure a wider range of temperatures (e.g. from –20 °C in a freezer to 400 °C in an oven).

HANDY TEMPERATURE METER DTM 1199
Monitors the temperature of materials, semi-finished products and finished products. Temperature ranges from –50 to 150 °C.
Power source: battery
Price code: 1
Physics Institute of Ho Chi Minh City, Vietnam

Thermometers are also manufactured by West Meters Ltd, UK.

38.6 pH measurement

FISHER ACCUMET (AP71) PH METERS
Easy to use, hand-held models come with a range of specifications and in various prices.
Power source: battery
Price code: 1/2
Fisher Scientific UK Ltd, UK

WATERPROOF POCKET PH METER
Easy to use, hand-held pH meter, completely waterproof and can be completely immersed in water. Has a positive push-button on/off switch. The easy to read display indicates when a stable pH value has been reached, ensuring that only accurate readings are taken. Calibration is automatic and can be performed easily, even by non-technical operators. The meter has an auto-switch-off feature (after 10 minutes) that prolongs the life of the batteries and gives up to 300 hours of use.
Power source: battery
Price code: 1
International Ripening Company, USA

GROCHECK METER PH, EC AND TDS
This combination model measures pH, electrical conductivity and total dissolved solids (for measuring fertilizer application). It is an easy to use, hand-held meter that is battery operated (total battery life is about 150 hours). It is supplied complete with probe, pH 7 buffer, 1.41 EC and 1500 ppm calibration solutions, a 9 V battery and instruction manual.
Power source: battery
Price code: 1
International Ripening Company, USA

COMBO WATERPROOF PH/EC/TDS TESTER
Designed for high accuracy pH, electrical conductivity, total dissolved solids and temperature measurements. Calibration and temperature compensation is automatic.
Power source: battery
Price code: 1
International Ripening Company, USA

38.7 Salt concentration measurement (salometer)

FISHERBRAND SALT WATER/BRINE UNIT
Hand-held two-scale unit that measures specific gravity and the concentration of salt in water (parts per thousand, ppt). Has automatic temperature compensation. The meter measures salinity from 0–100 ppt.
Price code: 1
Fisher Scientific UK Ltd, UK

39.0 Weighing equipment

It is important to have accurate weighing machines. Quite often, more than one weighing machine is required, one for weighing small quantities, say up to 1 kg in 10 g divisions and a separate one for measuring out larger quantities, say up to 30 kg in 100 g divisions.

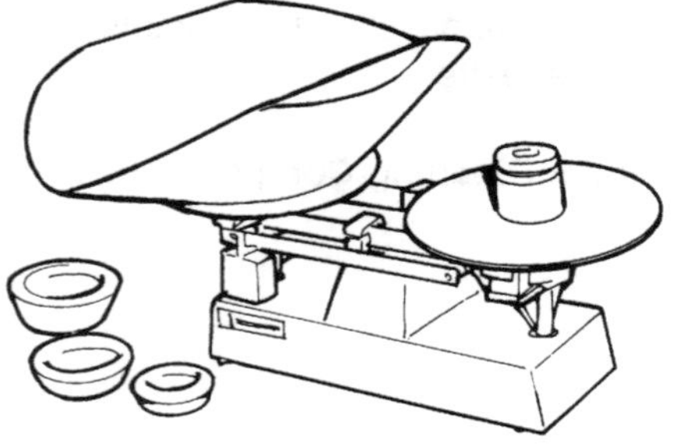

PLATFORM WEIGHERS
Platform scales used for the weighing of various products. Capacity: 250–1000 kg
Price code: 2/3
Alvan Blanch, UK

SEMI AUTOMATIC SACK WEIGHER
For weighing sacks of 10–100 kg.
Price code: 4
Alvan Blanch, UK

SMALL BAG WEIGHER (MINI ACCRAFILL)
A versatile automatic filling and weighing machine for handling a wide range of dry materials weighing between 1 and 10 kg. Interchangeable filling heads for all types of bags, cartons, tins, boxes, bottles etc. Continuous operation by alternate use of two filling heads. No power supply is needed.
Price code: 4
Alvan Blanch, UK

ELECTRONIC WEIGHING SYSTEMS
Including weighing scales, fluid filling machines and counting scales.
Essae-Teraoka Ltd, India

COUNTER WEIGHT SCALES
Capacity: 1–20 kg
Price code: 1
Gardners Corporation, India

COUNTER WEIGHT WEIGHING BALANCE
Capacity: 10 kg
Price code: 1
Gardners Corporation, India

WEIGHING MACHINE
Capacity: 300 kg
Price code: 1
Gardners Corporation, India

BAG WEIGHER
Premium Engineers Pvt Ltd, India

DIGITAL BATCH WEIGHING SCALE
The machine uses electronic control. Weighing process operates at two levels: big feed and small feed. Weight is sensed by load cell which produces a signal to operate pistons opening and closing two doors. Weighing range is 20–100 kg. Capacity: 6 tonnes/hour
Power source: electric
Price code: 3
Sai Gon Industrial Corporation (SINCO), Vietnam

WEIGHING SCALES
Scales are available in dial, electronics and platform.
Technoheat Ovens and Furnaces Pvt Ltd, India

39.1 Measuring equipment

WATER MEASURING TANK
Designed to supply water at a constant and predetermined temperature to the mixer. The tank is fitted with thermometers, control valves, two inlets, water level indicators, overflow and two-way outlet pipes for serving either of the dough kneader bowls.
Price code: 4
Baker Enterprises, India

40.0 Threshers

A range of threshing machines designed for use with different raw materials is available. Manual and powered versions in various sizes are available to suit most needs.

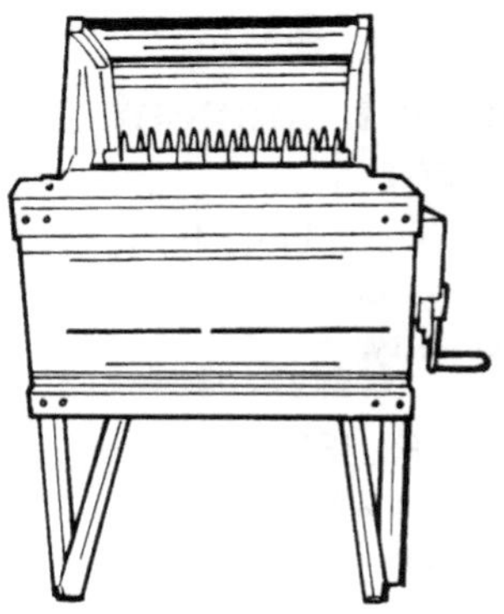

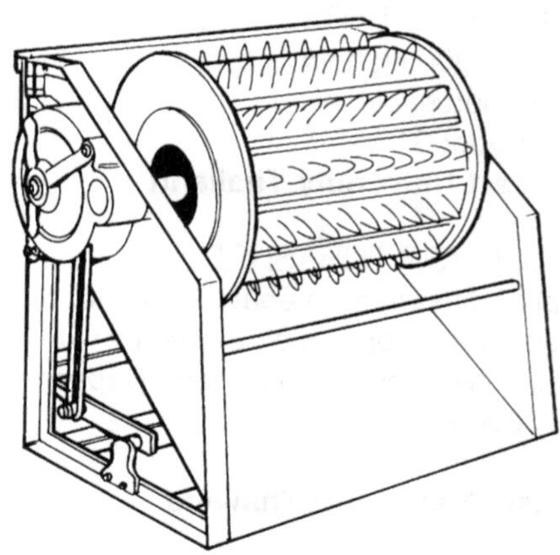

Powered

MIDGET MK 11A THRESHER
A low-cost portable machine which is highly adaptable and suits almost all crops. The grain sample will contain some chaff which can be removed by winnowing. Capacity: 500 kg/hour
Power source: diesel
Alvan Blanch, UK

SAMPLE THRESHER
Designed to thresh only the heads of the crop. After threshing the grain is metered into an adjustable airflow which removes the chaff. Capacity: 200 kg/hour
Power source: electric
Alvan Blanch, UK

NORTHSON JUNIOR GRAIN THRESHER
Portable machine with variable speed and simple adjustment to suit all types of seed.
Capacity: 200–1800 kg/hour
Power source: electric
G. North & Son (Pvt) Ltd, Zimbabwe

CORN HARVESTING MACHINE
Power source: diesel
Price code: 4
Kasetsart University, Thailand

PADDY RICE THRESHING MACHINE
Capacity: 300 kg/hour
Power source: diesel
Price code: 3
Kasetsart University, Thailand

MULTI CROP THRESHER
Simple adjustment to enable thresher to handle other crops. High efficiency with cleanliness of over 99 per cent. Capacity: 350–3000 kg/hour
Power source: diesel
Kunasin Machinery, Thailand

THRESHER
Capacity: 750 kg/hour
Power source: electric
Price code: 2
Kundasala Engineers, Sri Lanka

PADDY THRESHER
An engine-driven paddy thresher.
Power source: diesel
Mirpur Agricultural Workshop Training School (MAWTS), Bangladesh

HARVESTING AND SUGAR-JUICE SEPARATING MACHINE FOR SWEET SORGHUM
A prototype, low-cost harvesting machine for sweet sorghum, with a sugar-juice multi-pass extraction device and an integrated system for boiling the bagasse. Capacity: 0.5 hectares/hour
Pasquali Macchine Agricole srl, Italy

MULTICROP THRESHER
Capacity: 500–2000 kg/hour
Power source: electric/diesel
Rajan Universal Exports (Manufacturers) Pvt Ltd, India

PEPPER THRESHER
This machine removes berries from pepper spikes using a rotating drum with spikes. It has a 2 kW electric motor or diesel/kerosene engine. Capacity: 500–700 kg/hour
Power source: diesel/electric
Udaya Industries, Sri Lanka

JUTE STRIPPING MACHINE
Capacity: 500 kg/hour
Power source: electric
Price code: 2
Usa Patanasetagit Company Ltd, Thailand

HIGH SPEED MULTI CROP THRESHER
Capacity: 300–2000 kg/hour
Price code: 4
Votex Tropical, The Netherlands

Manual

HARVESTING MACHINE FOR SOYBEAN
Power source: manual
Price code: 2
Kasetsart University, Thailand

STRIPPER GATHERER SYSTEM
A basic pedestrian-controlled harvester which only strips and collects the grain. Separation of the straw, any re-threshing and cleaning need to be carried out as a separate operation.
Silsoe Research Institute, UK

STRIPPER THRESHER SYSTEM
A pedestrian-controlled stripper with a rotary straw separator/re-thresher. This machine strips the grain, separates it from unwanted straw and threshes any unthreshed panicles.
Silsoe Research Institute, UK

STRIPPER COMBINE SYSTEM
A pedestrian-controlled stripper plus rotary straw separator/re-thresher and cleaner.
Silsoe Research Institute, UK

41.0 Water treatment equipment

Equipment is available both for testing the quality of water and for treating water to improve quality.

WATER STERILISER
Water sterilizer with UV light.
Actini, France

WATER TREATMENT MACHINERY
Disinfects drinking/process water. Capacity: 75–150 000 litres/hour
Hitech Ultraviolet Pvt Ltd, India

42.0 Winnowers

A range of winnowing machines designed for use with different raw materials is available. Both manually operated and powered versions in a range of sizes are available to suit most needs.

PNEUMATIC ASPIRATOR
Used for 'air washing' of the material to remove all dust, lights and impurities which are lighter than the seed. Winnowing of harvested and threshed seeds. Capacity: 1000 kg/hour
Power source: diesel/electric
Price code: 1/2
Forsberg Agritech (India) Pvt Ltd, India

GORDON'S COCOA BEAN BREAKING AND WINNOWING SYSTEM
The system consists of an electrically driven cocoa breaker that cracks the whole bean – raw or roasted – and a pneumatic air-flow winnower that separates the broken shells from the nibs.
Power source: electric
John Gordon International, UK

CC1 WINNOWER
Separates the shells and husks of coffee and cocoa beans. Capacity: 15–45 kg/hour
Power source: electric
Price code: 2
John Gordon International, UK

NORTHSON JUNIOR WINNOWER
Simple adjustment for all crops. Capacity: 180 kg/hour
Power source: manual/electric
G. North & Son (Pvt) Ltd, Zimbabwe

43.0 Miscellaneous equipment

IRA SODA WATER MACHINE
Makes carbonated drinks using high pressure carbon dioxide. A pressure-regulating valve supplies CO_2 from a gas cylinder at an even pressure. A filler/crowner is operated by holding the bottle and pulling a lever. Capacity: 600 bottles per hour. A motorized version is also available.
Essence & Bottle Supply (India) Private Ltd, India

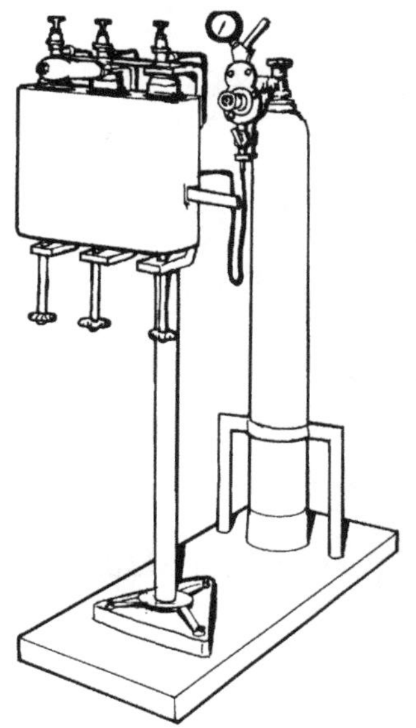

ESSENTIAL OIL DISTILLATION PLANT
Used to extract oil from a variety of crops using the method of distillation. Capacity: 130–420 kg/hour
Alvan Blanch, UK

SMALL-SCALE DISTILLATION EQUIPMENT
A range of glass distillation equipment, fragrance and flavour chemicals is available. Glass stills are constructed of 100% borosilicate glass to ensure a clean environment and the highest quality essential oils. The capacity of the stills ranges from 2 to 72 litres. The company also stocks accessories and replacement parts and carries out repairs.
Price code: 2
Floragenics Distillation Systems, USA

PORTABLE TABLE-TOP DISTILLER
This all-copper distillation unit is perfect for making essential oils by steam or water distillation. It is divided into three components – the retort, bird's beak and condenser. The capacity of the retort is 28.4 litres (7.5 gallons). The bird's beak attaches to the top of the retort and is equipped with a thermometer to monitor the temperature. The whole unit is also available in stainless steel.
Price code: 3
The Essential Oil Company, USA

FLORENTINE SEPARATOR
Florentine separator flasks are available as separate units and can be purchased independently of the distiller.
Price code: 2
The Essential Oil Company, USA

POULTRY DEFEATHERING MACHINE
Used for plucking chickens. Capacity: 100 chickens/hour
Power source: electric
Price code: 2
Bay Hiap Thai Company Ltd, Thailand

ELECTRIC GENERATOR
Useful for small-scale application.
Brouillon Process, France

STAINLESS STEEL TOPPED PREPARATION TABLE
Price code: 2
DISEG (Diseno Industrial y Servicios Generales), Peru

BOTTLING PLANT
Carbonating equipment suitable for making soda water and carbonated soft drinks.
Power source: electric/manual
Essence and Bottle Supply (India) Private Ltd, India

PILOT FERMENTER – FER 60/10
Used for the production of micro-organisms and/or metabolites in aerobic and anaerobic conditions by means of a batch, fed batch or continuous process. Capacity: 10–60 litres
Gauthier, France

ECM SYSTEM
Miniature manufacturing equipment for accurate tea research and development, quality assessment etc. Standard equipment includes Wither Cabinet, Ferment Cabinet, Fluid Bed Dryer, and Stalk Remover.
Teacraft Ltd, UK

Second-hand food-processing equipment is available from Scheitler, Argentina.

Manufacturers directory

Africa

Benin

Ere du Verseau
04 BP 310
Cotonou
Benin

Burkina Faso

AGCM (Atelier General de Construction Metallique)
Boulevard Chalons-en-Champagne
s/c 01 BP 265
Bobo-Dioulasso
Burkina Faso
Contact: M Guira Abdoul Karim

ASELEC (Atelier de Soudure et d'Electricité)
21–25 av. Kadiogo
BP 3171
Ouagadougou
Burkina Faso
Tel: 226 30 31 62

Atelier de Soudure Tout por Moulins
rue Vicens
BP 1163
Bobo-Dioulasso
Burkina Faso
Tel: 226 97 22 15
Contact: M Traore Karim

Atelier Konate B Boubacar
Rakieta
BP 178
Banfora
Burkina Faso
Contact: M Konate B Boubacar

CEAS-ATESTA (SAPE. SATA)
01 BP 5272
Ouagadougou
Burkina Faso
Tel: 226 20 23 93
Fax: 226 34 10 65
E-mail: ceas-rb@fasonet.bf
Contact: M Guisson Pierre

El Hadj Yaya Akewoula
av. William Pointy
BP 2428
Bobo-Dioulasso
Burkina Faso
Tel: 226 97 32 93
Contact: M Hadj Yaya Akewoula

Garage Outtara Bakari
Rakieta
BP 150
Banfora
Burkina Faso
Contact: M Ouattara Bakari

Guire Harouna (Etablissement)
à côté de la grande Mosque de Diarradougou
Bobo-Dioulasso
Burkina Faso
Contact: M G Harouna

Kabore Koutiga Jean et Frères (Etablissement)
01 BP 5349
Ouagadougou 07
Burkina Faso
Tel: 226 31 41 38
Contact: M Kabore Jean

Kabore Moussa et Frères (Etablissement)
en face de l'Assemblée des Dieux
BP 43
Koudougou
Burkina Faso
Tel: 226 44 07 12
Contact: M Kabore Moussa

Kinate et Frères (Etablissement)
BP 199
Bobo-Dioulasso
Burkina Faso
Tel: 226 98 26 323
Contact: M Konate Malamine

Nacanabo (Etablissement)
à côté du grand marche de Banfora
BP 225
Banfora
Burkina Faso
Contact: M Ouedraogo Adama

Ouedraogo Safiatou (Ets)
1620 av. de l'Unité à côté du bureau de douane de la gare
BP 3390
Bobo-Dioulasso
Burkina Faso
Tel: 226 97 06 18
Fax: 226 97 04 73
Contact: Mme Ouedraogo Safiatou

Sempore Issa et Frères (Etablissement)
av. de la Gare à côté de ST Wend Songoda s/c
BP 369
Koudougou
Burkina Faso
Tel: 226 44 10 23
Contact: M Sempore Issa

Société de Production et d'Exploitation de Matériel de Traitement du Karite
Secteur 2
BP 296
Koudougou
Burkina Faso
Tel: 226 44 03 58
Fax: 226 34 08 17
Contact: M Gadiaga Amadou

Soldev – Soleil et Développement
rue Vicens
BP 1539
Bobo-Dioulasso
Burkina Faso
Tel: 226 98 21 87
Contact: M Outtara

Tabsoba Salifou et Frères (Etablissement)
BP 1842
Ouagadougou
Burkina Faso
Tel: 30 83 53/31 34 72
Contact: M Tabsoba Salifou

Cameroon

Fabasem
BP 1485
Douala
Cameroon
Tel: 237 39 00 53

Helepac
BP 777
Yaounde
Cameroon
Tel: 237 30 61 19
Fax: 237 30 61 19

Outils Pour Les Communautes
PO Box 5946
Douala
Cameroon
Tel: 237 37 04 32
Fax: 237 37 04 02

Papier Plus
BP 4532
Douala
Cameroon
Tel: 237 37 50 12

Plasticam
BP 4071
Douala
Cameroon
Tel: 237 37 50 57

Printpak
BP 1005
Douala
Cameroon
Tel: 237 42 49 19
Fax: 237 42 59 49

Côte d'Ivoire

Société Ivoirienne de Technologie Tropicale
Port Bouet
04 BP 1137
Abidjan 04
Côte d'Ivoire
Tel: 225 21 23 68
Fax: 225 21 97 45

Ethiopia

Legio Aluminium and Electrical Bakery Industry
PO Box 2089
Addis Ababa
Ethiopia

Ghana

Agricultural Engineers Ltd
PO Box 3707
Accra
Ghana

Ghana Carton Boxes Mfg Ltd
PO Box 7676
Accra North
North Industries Area
Ghana
Tel: 233 21 228 492/228 408

John Kojo Arthur
University of Science and Technology
Kumasi
Ghana

Kaddai Engineering
PO Box 2268
Kumasi
Ghana

Plastics Packaging Products Ltd
PO Box 483
Accra
North Industrial Area
Ghana
Tel: 233 21 228 882
Fax: 233 21 222 72

SSIS Enterprises
Kumasi
Ghana

Technology Consultancy Centre
University of Science and Technology
Kumasi
Ghana
Tel: 233 51 60297
Fax: 233 51 60137

Top Industrial Packaging Products Ltd
PO Box 5274
Accra North
Kaneshie Industrial Area
Ghana
Tel: 233 21 225 973

Tropical Glass Co. Ltd
Takoradi
Tarkwa
PO Box 8
Aboso
Ghana
Tel: 233 21 362 41 68

WA ITTU
Gratis Project
Ministry of Environment, Science and Technology
PO Box 226
WA
Ghana

Kenya

ApproTEC – Appropriate Technologies for Enterprise Creation
PO Box 64142
Nairobi
Kenya
Tel: 254 2 783046
E-mail: info@approtec.org
Web: www.approtec.org/

Mali

Sada-SA
BP 1110
Bamako
Mali
Tel: 223 21 37 51/21 49 32
Fax: 223 21 26 52

Mozambique

Agro Alfa
Avenue Angola no. 2475
CP 4209
Maputo
Mozambique
Tel: 258 1 465911/465258

Nigeria

Nigerian Oil Mills Ltd
PO Box 264
Atta
Owerri
Imo-State
Nigeria

Sierra Leone

Department of Agriculture, Forestry and the Environment
PO Box 18
Magburaka
Tonkolili District
Sierra Leone

South Africa

Bee Keeper's Supermarket
38 Milner Road
Maitland
Cape Town 7450
South Africa
Tel: 27 21 511 4567
Fax: 27 21 511 9962

Beeswax Comb Foundation
340 Boom Street
Pietermaritzburg 3201
South Africa
Tel: 27 331 451016

Highveld Honey Farms
PO Box 11079
Rynfield Tvl.
1514
South Africa
Tel: 27 11 849 1990

Stilfontein Apiaries
PO Box 1145
Stilfontein 2550
South Africa
Tel: 27 18 484 5694
Fax: 27 18 484 5694

Woodlands Beehives
PO Box 17770
Pretoria North 0116
South Africa
Tel: 27 12 561 1640
Fax: 27 12 561 1640

Tanzania

Centre for Agricultural Mechanisation and Rural Technology (CAMARTEC)
PO Box 764
Arusha
Tanzania
Tel: 255 57 8250

John Gordon International
East African Regional Office
Tanzania
Tel: 255 632 2148
Fax: 255 632 2148

Karume Technical College
PO Box 467
Zanzibar
Tanzania

Space Engineering Ltd
Nelson Mandela Road
Dar Es Salaam
Tanzania

TEMDO (Tanzania Engineering and Manufacturing Design Organisation)
PO Box 6111
Arusha
Tanzania
Tel: 255 57 8058
Fax: 255 57 8318

Uganda

Northern Uganda Manufacturers Association
PO Box 296
Lira
Uganda

Uganda Small Scale Industries Association
PO Box 7725
Kampala
Uganda

Zimbabwe

G. North & Son (Pvt) Ltd
PO Box St. 111
Southerton
Harare
Zimbabwe
Tel: 263 4 63717/9

H.C. Bell & Son Engineers (Pvt) Ltd
24 Glasgow Road
PO Box 701
Mutare
Zimbabwe
Tel: 263 20 63094/63018
Fax: 263 20 62535

Asia

Bangladesh

Intermediate Technology Development Group Bangladesh
GPO Box 3881
Dhaka 1000
Bangladesh
Tel: 880 2 811 1934/911 0060/912 3671
Fax: 880 2 811 3134
E-mail: itdg@bdmail.net

Mark Industries (Pvt) Ltd
348/1 Dilu Road
Mokbazar
Dhaka-1000
Bangladesh
Tel: 880 2 9331778/835629/835578
Fax: 880 2 842049
E-mail: markind@citechco.net

Mirpur Agricultural Workshop & Training School (MAWTS)
Pallabi
Mirpur-12
Dhaka
Bangladesh
Tel: 880 2 9002544/803810/9002493
Fax: 880 2 801107
E-mail: mawts@bdonline.com

Modern Erection
223-B Tejgaon Industrial Area
Dhaka 1208
Bangladesh
Tel: 880 2 601903/600552/9882468
Fax: 880 2 872304

India

Acufil Machines
SF 120/2
Vazhiyampalayam Road
Kalapatty
Coimbatore – 641 035
India
Tel: 91 422 866108/866255
Fax: 91 422 572640
E-mail: gondalu@yahoo.com

Admir Enterprises
Plot No. 1/E-4 Shivaji Nagar
Govandi
Mumbai – 400 043
India
Tel: 91 22 556 8446/551 8182/767 2067/763 5144
Fax: 91 22 556 8446

Agaram Industries
126 Nelson Road
Aminjikarai
Chennai – 600 029
Tamil Nadu
India

Agro Industrial Agency
Near Malaviya Vadi
Gondal Road
Rajkot – 360 002
India
Tel: 91 281 461134/462079/451214
Fax: 91 281 461770
E-mail: jagdishindia@hotmail.com

Aifso Industrial Equipments Co.
B/13, Veena-Beena Apts
P Thakrey Marg
Sewri (W)
Mumbai – 400 015
India
Tel: 91 22 414 02 92/416 88 35
Fax: 91 22 416 8835

Ajay & Abhay (P) Ltd
B-84/1, Okhla Industrial Area – 2
New Delhi – 110020
India
Tel: 91 11 683 1215/6539
Fax: 91 11 683 0190
E-mail: jaipuria@vsnl.com

Alven Foodpro Systems (P) Ltd
24 Gold Fields Plaza
45 Sassoon Road
Pune – 411 001
India
Tel: 91 212 628516
Fax: 91 212 621534
E-mail: invendia@jwbbs.com

AMI Engineering
Station Road (opposite Veena Cinema)
Patna – 800001
Bihar
India
Tel: 91 612 224274/262847

Apple
14A Nisar Building
Sleater Road
Grant Road
West Mumbai – 400 007
India
Tel: 91 22 301 2193
Fax: 91 22 301 5931
E-mail: bakapple@bom5.vsnl.net.in

Autopack Machines Pvt Ltd
101-C, Poonam Chambers
A' Wing 1st Floor
Dr Annie Besant Road, Worli
Mumbai – 400 018
India
Tel: 91 22 493 4406/497 4800/492 4806
Fax: 91 22 4964926
E-mail: autopack@bom3.vsnl.net.in

Azad Engineering Company
C-83
Bulandshahar Road
Industrial Area
Ghaziabad – 201009 (UP)
India
Tel: 91 575 700708/730122
Fax: 91 575 702816

Bajaj Maschinen Pvt Ltd
C-582
New Friends Colony
New Delhi – 110 065
India
Tel: 91 11 226 5362/2191461
Fax: 91 11 217 1151
E-mail: bajaj@del3.vsnl.net.in

Baker Enterprises
23 Bhera Enclave
Near Peera Garhi
New Delhi – 110087
India
Tel: 91 11 558 6238/567 5265
Fax: 91 11 558 6150
E-mail: bredo@giasdl01.vsnl.net.in

Bhavani Sales Corporation
Plot No. 2/1, Phase II
GIDC
Vatva
Ahmedabad – 382 445
India
Tel: 91 79 583 1346/589 3253
Fax: 91 79 583 5885/583 1346
E-mail: labeling@ad1.vsnl.net.in

Bijoy Engineers
Mini Industrial Estate
Arimpur
Thrissur – 680 620
Kerala
India
Tel: 91 487 630451
Fax: 91 487 421562

Bombay Engineering Industry
R. No. 6 (Extn) Sevantibai Bhavan
Chimatpada
Marol Naka, Andheri (East)
Mumbai – 400 059
India
Tel: 91 22 836 9368/91 22 821 5795
Fax: 91 22 8369368

Bombay Engineering Works
1 Navyug Industrial Estate
185 Tokersey Jivraj Road
Opposite Swan Mill, Sewree (W)
Mumbai
India
Tel: 91 22 413 7094/413 5959
Fax: 91 22 413 5828
E-mail: bomeng@bom3.vsnl.net.in

Bombay Industrial Engineers
430 Hind Rajasthan Chambers
D.S. Phalke Road
Dadar (C. Rly.)
Mumbai – 400 014
India
Tel: 91 22 411 39 99/411 42 75
Fax: 91 22 620 1914

B Sen Barry & Co.
65/11 Rohtak Road
Karol Bagh
New Delhi – 110005
India
Tel: 91 11 572 3553/572 1105

Cantech Machines
13 Vora Bhuwan
Maheshwari Udyan, above Dena Bank
Matunga (C. Rly.)
Mumbai – 400 019
India
Tel: 91 22 409 6086/409 6853
Fax: 91 22 409 6086
E-mail: dpandya@bom3.vsnl.net.in

Capitol Engineering Works
Plot No. C-30, Road No. 16
Wagle Industrial Estate
Thane 400 604
Maharashtra
India
Tel: 91 22 532 2195/532 2070
Fax: 91 22 532 3521

Central Food Technological Research Institute
Cheluvamba Mansion
KRS Road
PO Food Technology
Mysore – 570 013
India
Tel: 91 821 22660
Fax: 91 821 517233
E-mail: cftri@ttm.ernet.in

Central Institute of Agricultural Engineering
Nabi Bagh
Berasia Road
Bhopal – 462 038
Madhya Pradesh
India
Tel: 91 755 530980–87
Fax: 91 755 534016
E-mail: ciae@x400.nicgw.nic.in

Cip Machineries Pvt Ltd
10 Umlya Estate
Bharat Party Plot, NH – 8 Road
Amraiwadi Road
Ahmedabad – 380 026
India
Tel: 91 79 289 1410/287 5410
Fax: 91 79 289 1410

Dairy Udyog
C-230, Ghatkopar Industrial Estate
LBS Marg
Ghatkopar (West)
Mumbai – 400 086
India
Tel: 91 22 517 1636/517 1960
Fax: 91 22 517 0878
E-mail: jipun@vsnl.com

DIW Precision Engineering Works
Vadkun Road
Dahanu Road
Thane 401 602
Maharashtra
India

Eastend Engineering Company
173/1 Gopal Lal Thakur Road
Calcutta – 700 035
India
Tel: 91 33 577 3416/6324
Fax: 91 33 556 6710/160

Essae-Teraoka Ltd
377/22, 6th Cross Wilson Garden
Bangalore – 560027
India
Tel: 91 80 2216185/2241165
Fax: 91 80 2225920

Essence and Bottle Supply (India) Private Ltd
14 Radha Bazar Street
PO Box 372
Calcutta – 700 001
India
Fax: 91 33 225 5358

Forsberg Agritech (India) Pvt Ltd
123 GIDC Estate
Makarpura
Baroda – 390 010
India
Tel: 91 265 645752
Fax: 91 265 641683
E-mail: Forsagri@ad1.vsnl.net.in

Gardners Corporation
158 Golf Links
New Delhi – 110003
India
Tel: 91 11 334 4287/336 3640
Fax: 91 11 371 7179

Geeta Food Engineering
Plot No. C-7/1 TTC Area
Pawana MIDC, Thane Belapur Road
Behind Savita Chemicals Ltd
Navi Mumbai – 400 705
India
Tel: 91 22 782 6626/766 2098
Fax: 91 22 782 6337

Goldin (India) Equipment (Pvt) Ltd
F-29, BIDC Industrial Estate
Gorwa
Vadodara – 390 016
Gujarat
India
Tel: 91 265 380 168/380 461
Fax: 91 265 380 168/380 461
E-mail: goldin@EQUIP.XEEBDQ.xeemail.com

Goma Engineering Pvt Ltd
Majiwada
Behind Universal Petrol Pump
Thane – 400 601
Maharashtra
India
Tel: 91 22 534 6436/534 0875
Fax: 91 22 533 3634/3632

Gurdeep Packaging Machines
Harichand Mill Compound
LBS Marg
Vikhroli
Mumbai – 400 079
India
Tel: 91 22 578 3521/577 2846/579 5982
Fax: 91 22 577 2846

Hariom Industries
Dhebar Road (South)
Atika Industrial Area
Near Jaidev Foundry
Rajkot – 360 002
India
Tel: 91 281 363620/371745

Hitech Ultraviolet Pvt Ltd
PO Box No. 8356
Grace Plaza
Jogeshwari (W)
Mumbai – 400 102
India
Tel: 91 22 679 0610/679 4611
Fax: 91 22 679 4337

Industrial Refrigeration Pvt Ltd
901 Maker Chambers V
Nariman Point
Mumbai – 400 021
India
Tel: 91 22 204 1183/5/9, 287 2363/79
Fax: 91 22 204 4944
E-mail: alistate.finance@gems.vsnl.net.in

John Fowler (India) Ltd
Sarjapur Road
Bangalore – 560 034
India
Tel: 91 80 5530026
Fax: 91 80 5533228
E-mail: jfil@blr.vsnl.net.in

Jwala Engineering Company
12 Surve Industrial Estate
Sonawala Cross Road No. 1
Goregaon (East)
Mumbai – 400 063
India
Tel: 91 22 874 0279
Fax: 91 22 876 8768

Kaps Engineers
831 GIDC
Makarpura
Vadodara – 390 010
India
Tel: 91 265 644692/640785/644407
Fax: 91 265 643178/642185

Karishma Instruments Pvt Ltd
C-16, Rajkumar Apts.
17th Road
Santacruz (West)
Mumbai – 400 054
India
Tel: 91 22 649 2493
Fax: 91 22 646337
E-mail: karishma@bom3.vsnl.net.in

Machin Fabrik
B-12/13, Arjun Centre
B S Devishi Marg
Govandi
Mumbai – 400 088
India
Tel: 91 22 5555596/5560947
Fax: 91 22 5560569

MMM Buxabhoy & Co.
140 Sarang Street
1st Floor
Near Crawford Market
Mumbai
India
Tel: 91 22 344 2902
Fax: 91 22 345 2532
E-mail: yusufs@glasbm01.vsnl.net.in

M/S Mangal Engineering Works
Factory Area
Patiala – 147001
India
Tel: 91 175 215180/214702/212652
Fax: 91 175 212652

M/S Wonderpack Industries (P) Ltd
PO Box No. 29127
321 TV Industrial Estate
SK Ahire Marg, Worli
Mumbai – 400 025
India
Tel: 91 22 493 4736/493 9580
Fax: 91 22 493 8796/436 3960

Narangs Corporation
P-25, Connaught Place
New Delhi – 110 001
India
Tel: 91 11 336 3547
Fax: 91 11 374 6705

North Eastern Regional Institute of Science
& Technology (NERIST)
Department of Agricultural Engineering
Nirjuli 791109
Arunchal Pradesh
India

Orbit Equipments Pvt Ltd
Block No. 1
Venkat Reddy Complex
Tarbund X Roads
Secunderabad – 560 009
India
Tel: 91 40 817296
Fax: 91 40 813877

Pharmaco Machines
2 Yashwant Appt
RS Road, Pramila Nivas Compound
Chendani
Thane – 400 602
India
Tel: 91 22 542 8478
Fax: 91 22 544 8469

Pharmaconcept
203 Malwa, 2nd Floor
E.S. Patanwala Compound
LBS Marg, Ghatkopar (West)
Mumbai – 400086
India
Tel: 91 22 500 3400/500 0178
Fax: 91 22 500 1358

Premium Engineers Pvt Ltd
603 Chinubhai Centre
Nehru Bridge Corner
Ashram Road
Ahmedabad – 380009
India
Tel: 91 79 657 9293/5987
Fax: 91 79 657 7197

Prima Engineering Industries
T-113 MIDC
Bhosari
Pune – 411026
India
Tel: 91 212 790621
Fax: 91 212 790294

Rajan Universal Exports (Manufacturers) Pvt Ltd
Post Bag No. 250
162 Linghi Chetty Street
Chennai – 600 001
India
Tel: 91 44 5341711/5340731/5340751/5340356
Fax: 91 44 5342323

Rank and Company
A-95/3
Wazirpur Industrial Estate
Delhi – 110 052
India
Tel: 91 11 7456 101/2/3/4
Fax: 91 11 7234 126/7433905
E-mail: rank@poboxes.com

Rieco Industries Ltd
1162/2 Shivajinagar
Behind Observatory
Pune – 411 005
India
Tel: 91 212 325384/325215
Fax: 91 212 323229

RPM Engineers (India) Ltd
44A NP Developed Plots
Thiru-vi-ka Industrial Estate
Ekkattuthangal
Chennai – 600 097
India
Tel: 91 44 234 6563/234 6679
Fax: 91 44 232 1639

Shirsat Electronics
12/82 Govind Nagar
Sodawala Lane
Borivali (W)
Mumbai – 400 092
India
Tel: 91 22 801 0250

Solar Arks
G-111 MIDC Gokul Shirgaon
Kolhapur – 416 234
Maharashtra
India
Tel: 91 231 672486
Fax: 91 231 672413

Sree Manjunatha Roller Flour Mills (P) Ltd
100/B Belagola Industrial Area
KRS Road
Mysore – 570 016
India
Tel: 91 821 510186/518361/518362
Fax: 91 821 513840

Sridevi Packing Industries
No. B-1 & B-2
Hebbal Industrial Area
Mysore – 570016
India
Tel: 91 821 512496/514958
Fax: 91 821 512496

Sri Murugan Industries
Site No. 37-B
II Stage
Industrial Suburb
Mysore – 570 008
India
Tel: 91 821 515993

Sri Rajalakshmi Commercial Kitchen Equipment
No. 57 (old No. 30/1)
Silver Jubilee Park Road
Bangalore – 560 002
India
Tel: 91 80 222 1054/223 9738
Fax: 91 80 222 2047

Sri Venkateswara Industries
C-37 Industrial Estate
Yadavagiri
Mysore – 570 020
India
Tel: 91 821 514828/443310

Sunray Industries
10 Paramahamsa Road
1st Main
Yadavagiri
Mysore – 570 020
India
Tel: 91 821 514324/515946
Fax: 91 821 515966

Techno Equipments
Saraswati Sadan
1st Floor
31 Parekh Street
Mumbai – 400 004
India
Tel: 91 22 385 1258

Technoheat Ovens and Furnaces Pvt Ltd
No. L1, 7th Cross
1st Stage
Peenya Industrial Estate
Bangalore – 560 058
India
Tel: 91 80 839 2890/837 0899
Fax: 91 80 837 1418

Tinytech Plants
Near Bhaktinagar Station
Tagore Road
Rajkot – 360 002
India
Tel: 91 281 480166/451086
Fax: 91 281 467552/453231
E-mail: tinytech@tinytechindia.com

Wintech Taparia Ltd
25/1 YN Road
3rd Floor, Shreesh Chambers
Indore – 452 003
Madhya Pradesh
India
Tel: 91 731 433 950/534 586
Fax: 91 731 430527
E-mail: sales@wintechtaparia.com

Indonesia

Natural Technik
J1 Palmerah Barat, 48A
Jakarta
Indonesia
Tel: 62 21 548 1345/2084

Japan

Arai Machinery Corporation
2–7–19 Okata
Atsugi-Shi
Kanagawa-Ken 243–0021
Japan
Tel: 81 46 2270461
Fax: 81 46 2270463

Philippines

Almeda Food Machinery Corporation
1337 Dalaga Street
Tondo
Manila
The Philippines
Tel: 63 2 253 6591/6598/6601
Fax: 63 2 635 3005

John Gordon International
Far Eastern Regional Office
Manila
The Philippines
Tel: 632 844 3079
Fax: 632 844 1012

Well Done Metal Industries
2438 Juan Luna Street
Galalangin
Tondo
Manila 1012
The Philippines
Tel: 63 2 253 95 35/41
Fax: 63 2 253 95 35

Sri Lanka

Ashoka Industries
Kirama
Walgammulla
Sri Lanka
Tel: 94 71 764725

Coconut Development Authority
54 Nawala Road
Colombo 5
Sri Lanka
Tel: 94 1 502503/4
Fax: 94 1 508729

Kundasala Engineers
Digana Road
Kundasala
Kandy
Sri Lanka
Tel: 94 8 420482

Odiris Engineering Company
43 Dutugemumu Street
Pamankada
Dehiwala
Sri Lanka
Tel: 94 1 811063
Fax: 94 1 811063

Silva Industries
337/8 Polgahpitya
Ambagaha Junction
Mulleriyawa New Town
Sri Lanka
Tel: 94 1 572051
Fax: 94 1 572051

Sri Lanka Cashew Corporation
349 Galle Road
Colombo 03
Sri Lanka
Tel: 94 1 576057/576054
Fax: 94 1 577267
E-mail: cashewco@sltnet.lk

Udaya Industries
Uda Aludeniya
Weligalla
Gampola
Sri Lanka
Tel: 94 8 388586
Fax: 94 8 388909

Thailand

Amorn Loharyon
585/12 Soi Kingchan
Chan Road
Yannawa
Bangkok
Thailand
Tel: 66 2 2116234

Banyong Engineering
94 Moo 4 Sukhaphibaon No. 2 Rd
Industrial Estate Bangchan
Bankapi
Thailand
Tel: 66 2 5179215–9

Bay Hiap Thai Company Ltd
Wernakorn Kasem
Bangkok
Thailand
Tel: 66 2 2212432/2213670

Charoenchai Company Ltd
7–9 Yaowarat
Near Panuphon Bridge
Bangkok
Thailand
Tel: 66 2 2223668/2222743

Eamseng
5–7 Wernakornkasem
Charoenkrung Road
Bangkok
Thailand
Tel: 66 2 2213451/2234548

Flower Food Company
91/38 Ram Indra Road
Soi 10
Km 4
Bangkok
Thailand
Tel: 66 2 5212203/5523420

Kasem Kanchang
Thonburi
Bangkok
Thailand

Kasetsart University
Bangkaen Campus
50 Paholyothin Road
Bangkaen
Bangkok 10900
Thailand
Tel: 66 2 5790572/5790113/5614484

Kongsonglee Kanchang
Charoensanitwongse Road
Bangkok
Thailand
Tel: 66 2 411 1462

Krungthep Chanya
Charansanitwonge
Bangphlad
Bangkok
Thailand
Tel: 66 2 424063 5/4244714

KSL Engineering
275/1 Soi Thaweewatana Satupradit
Yannawa
Bangkok
Thailand
Tel: 66 2 2118497

Kunasin Machinery
34 Soi Ladphrod, 125 Ladphrod Road
Wangtonglong
Bangkaip
Bangkok
Thailand
Tel: 66 55 642119/641653

Lim Chieng Seng Ltd
92–94 Sawanwithee Road
Amphoe Muang
Nakornsawan 60000
Thailand
Tel: 66 56 221197/221765

Narongkanchang
783/3 Thadindaeng Rd
Krongsarn
Bangkok
Thailand

Ruang Thong Machinery Ltd
213/5–6 Soi Tavevathana
Sathupradit Road
Yannava
Bangkok
Thailand
Tel: 66 2 2117744/2359919/2335015
Fax: 66 2 2385279

Sahathai Factory
Chan Road,
Yannawa
Bangkok
Thailand
Tel: 66 2 2113666

Sahathai Sathorn
Bangkok
Thailand

Saw Charoenchai
553 New Boripat Road
Sapang Lek
Werngnakorn Kasem Charoenkrung
Bangkok
Thailand
Tel: 66 2 2214590/2259613

Thaworn Kanchang
Suksawat Rd, Soi 26
Bangpakod
Radburana
Bangkok
Thailand
Tel: 66 2 4273171/4273010

Usa Patanasetagit Company Ltd
56/7 Prachacheun Toonsonghong Bangkaen
Bangkok 10211
Thailand
Tel: 66 2 5892221/5890935

Vietnam

ALFA Technology Transfer Centre
301 Cach Mang Thang 8
Tan Binh District
Ho Chi Minh City
Vietnam
Tel: 848 9700868
Fax: 848 8640252
Contact: Mr Le Thuong

Anh Tuan Mechanical Cooperative
34 Hang Ma Street
Hanoi
Vietnam

Bui Van Thanh Private Company
71B Hang Ma Street
Hanoi
Vietnam
Tel: 844 8281 768

Dinh Chi Thang
50 Hang Ma Street
Hanoi
Vietnam
Tel: 844 8280680

Dinh Van Bay
24 Loren Street
Hanoi
Vietnam
Tel: 844 8257464

Doan Binh Mechanical Cooperative
52 Hang Ma Street
Hanoi
Vietnam
Tel: 844 9432427

Duc Huan Mechanical Cooperative
58 Hang Ma Street
Hanoi
Vietnam
Tel: 844 9230950

Mechanical Engineering Faculty,
University of Agriculture
Thu Duc District
Ho Chi Minh City
Vietnam
Tel: 848 8960721
Fax: 848 8960713
Contact: Professor Nguyen Nhu Nam

Physics Institute of Ho Chi Minh City
1 Mac Dinh Chi
District 1
Ho Chi Minh City
Vietnam
Tel: 848 8224890
Fax: 848 8234133
Contact: Mr Duong Minh Tri

Research Institute for Agricultural Machinery
Km 9.5 Nguyen Trai Road
Hanoi
Vietnam
Tel: 844 8547363
E-mail: vmnn@hn.vnn.vn
Contact: Professor Nguyen Tuong Van

Sai Gon Industrial Corporation (SINCO)
63–65 Tran Hung Dao
District 1
Ho Chi Minh City
Vietnam
Tel: 848 8367761
Fax: 848 8368809
Contact: Mr Bui Quoc An

Technology & Equipment Development
Centre (LIDUTA)
360 Bis Ben Van Don St
District 4
Ho Chi Minh City
Vietnam
Tel: 848 9400906
Fax: 848 9400906
Contact: Mr Vo Hoang Liet

Van Duc Cooperative
98 Ton Duc Thang Street
Hanoi
Vietnam
Tel: 844 8434906

Xuan Kien Private Company
60 Ly Thuong Kiet Street
Hoan Kiem District
Hanoi
Vietnam
Tel: 844 8268640
Fax: 844 8223435
Contact: Eng. Bui Ngoc Huyen

Europe

Austria

J. Wick Filling Machines
Caaerberg Str. 100
A-1100 Vienna
Austria
Tel: 43 1 682317
Fax: 43 1 682316

Belgium

Ets Deklerck
Place Lehon 14
B-1030 Brussels
Belgium
Tel: 32 2 215 54 87
Fax: 32 2 216 47 94

Denmark

APV Unit Systems
Pasteursvej
DK-8600 Silkeborg
Denmark
Tel: 45 8922 8922
Fax: 45 8922 8901

Gerstenberg & Agger A/S
19 Frydendalsvej
DK-1809 Frederiksberg C
Copenhagen
Denmark
Tel: 45 31 31 28 39
Fax: 45 31 31 30 25

Swienty A/S
Hortoftvej 16
Ragebol
DK 6400 Sonderborg
Denmark
Tel: 45 74 48 69 69
Fax: 45 74 48 80 01
E-mail: swienty@aof.dk
Web: www.swienty.dk

France

Actini
Parc de Montigny
Maxilly-sur-Leman
75500 Evian-les-Bains
France
Tel: 33 4 50 70 74 74
Fax: 33 4 50 70 74 75

Actini International
BP 180
9 avenue du Gal de Gaulle
74202 Thonon Cedex
France
Tel: 33 4 50 71 73 93
Fax: 33 4 50 26 39 08

Altech
rue des Cordeliers
5200 Embrun
France
Tel: 33 4 92 43 21 90

Biaugeaud Henri
45 avenue Aristide Briand
BP 17
94111 Arcueil Cedex
France
Tel: 33 1 42 53 77 40
Fax: 33 1 42 53 11 26

Brouillon Process
BP 15
47180 Sainte-Bazeille
France
Tel: 33 5 53 20 98 00
Fax: 33 5 53 64 72 80

Capic
ZI de l'Hyppodrome
BP 613
29551 Quimper Cedex 9
France
Tel: 33 2 98 82 77 00

Elecrem
24 rue Gambetta
BP 45
92174 Vanves Cedex Mr Tourret
France
Tel: 33 1 46 42 14 14

Electra
47170 Poudenas
France
Tel: 33 5 65 73 55
Fax: 33 5 53 97 33 05

Elimeca SA
14 a 22 rue Rabutin
1140 Thoissey
France
Tel: 33 4 74 69 76 90
Fax: 33 4 74 69 72 75

Gauthier
Parc Scientifique Agropolis
34397 Montpellier
Cedex 5
France
Tel: 33 4 67 61 11 56
Fax: 33 4 67 54 73 90

Geere SA
51 route d'Orleans
45410 Artenay
France
Tel: 33 38 80 44 80

Gilson Père et Fils
74150 Hauteville-sur-Fier
France
Tel: 33 4 50 60 50 16
Fax: 33 4 50 60 52 31

La Mécanique Moderne
31 rue Saint-Michel
BP 103
62002 Arras
France
Tel: 33 3 21 55 36 00
Fax: 33 3 21 21 04 34

Marot
Agro-alliance
BP 72
60304 Senlis Cedex
France
Tel: 33 3 44 60 03 00
Fax: 33 3 44 53 21 60

Olier Ingenierie
ZI Le Brezet
31 rue des Frères Lumière
63100 Clermond Ferrand
France

Robot Coupe
12 avenue Gal Leclerc
BP 134
71303 Montceau-les-Mines
France
Tel: 33 3 85 58 80 80/66 66

Samap
1 rue du Moulin
69280 Andolsheim
France

Société Pontchatelaine Equipement Cidricole
32bis rue Maurice Sambron
BP 23
44160 Pontchateau
France
Tel: 33 2 40 88 11 39
Fax: 33 2 40 01 66 82

Germany

Gebruder Lodige Maschinenbau GmbH
Postfach 2050
D-33050 Paderborn
Germany
Tel: 49 5251 309–0
Fax: 49 5251 309–123

Graze
Staffelstr. 5
71384 Weinstadt (Endersbach)
Germany
Tel: 49 71 51 96 92 30
Fax: 49 71 51 96 92 33

IBG Monforts GmbH & Co.
Postfach 200853
D-41208 Münchengladbach 2
Germany
Tel: 49 2166 86820
Fax: 49 2166 868244

Innotech
Ingenieursgesellschaft mbH
Brandenburger Strasse 2
D-71229
Leonberg
Germany
Tel: 49 7152 76101

Italy

Pasquali Macchine Agricole srl
Via Nuova 3
I-50041-Calenzano
Florence
Italy
Tel: 39 55 8879541
Fax: 39 55 8877746

Lithuania

Joint-Stock Company Radviliskis Machine Factory
Vytauto 3
LT-5120 Radviliskis
Lithuania
Tel: 370 92 52379/53248
Fax: 370 92 53439/51382

The Netherlands

C. van 't Riet Zuiveltechnologie BV
Dorpsstraat 25
Postbus 1008
2445 AJ Aarlanderveen
The Netherlands
Tel: 31 172 5 71304
Fax: 31 172 5 73406
E-mail: info@rietdairy.nl
Web: www.rietdairy.nl/

Small Scale Dairy Technology Group
Wildforster 37
6713 KA Ede
The Netherlands
Tel: 31 8380 24235

Votex Tropical
Vogelenzang Andelst BV
Wageningestraat 30
Andelst
NL-6673DD
The Netherlands
Tel: 31 488 469500
Fax: 31 488 454041

Spain

Talsabell SA
Pol. Ind. V. Salud, 8
E-46950 Xirivella
Valencia
Spain
Tel: 34 96 3591111
Fax: 34 96 3599828
E-mail: talsa@nexo.es

Switzerland

Novo Nordisk Ferment Ltd
Neumatt
4243 Dittingen
Switzerland
Tel: 41 61 7656111
Fax: 41 61 7656333

Ukraine

Odessa State Academy of Food Technology
112 Kanatna Street
Odessa
270039
Ukraine
Tel: 380 48 225 3284
Fax: 380 48 225 3284
E-mail: george@osaft.odessa.ua

United Kingdom

AFOS Ltd
Springfield Way
Anlaby
Hull
HU10 6RL
UK
Tel: 44 1482 352152
Fax: 44 1482 565265
E-mail: food@afos.ltd.uk
Web: www.afos.ltd.uk

Alvan Blanch
Chelworth
Malmesbury
Wiltshire
SN16 9SG
UK
Tel: 44 1666 577333
Fax: 44 1666 577339
E-mail: enquiries@alvanblanch.co.uk
Web: www.alvanblanch.co.uk

Armfield Ltd
Bridge House
West Street
Ringwood
Hampshire BH24 1DY
UK
Tel: 44 1425 478781
Fax: 44 1425 470916
E-mail: sales@armfield.co.uk

Charles Wait (Process Plant) Ltd
151 Fylde Road
Southport
Merseyside
PR9 9XP
UK
Tel: 44 1704 211273
Fax: 44 1704 225875
E-mail: sales@cwpp.co.uk

Christy
Foxhills Industrial Estate
Scunthorpe
North Lincolnshire
DN15 8QW
UK
Tel: 44 1724 280514
Fax: 44 1724 282123

Crypto Peerless Ltd
Bordesley Green Road
Birmingham
B9 4UA
UK

Danfoss Ltd
Perivale Industrial Park
Horsenden Lane South
Greenford
Middlesex UB6 7QE
UK
Tel: 44 208 991 7000
Fax: 44 208 991 7149

Denlab International (UK) Ltd
Hillcrest House
4 Market Hill
Maldon
Essex CM9 4PZ
UK
Tel: 44 1621 858944
Fax: 44 1621 857733
E-mail: admin@johngordon.co.uk

Fisher Scientific UK Ltd
Bishop Meadow Road
Loughborough
Leicestershire
LE11 5RG
UK
Tel: 44 1509 231166
Fax: 44 1509 231893
E-mail: fisher@fisher.co.uk
Web: www.fisher.co.uk

Fullwood Ltd
Grange Road
Ellesmere
Shropshire
SY12 9DF
UK
Tel: 44 1691 622391
Fax: 44 1691 622355

Glen Creston Ltd
16 Dalston Gardens
Stanmore
Middlesex
HA7 1BU
UK
Tel: 44 208 206 0123
Fax: 44 208 206 2452

HRP Focus Ltd
PO Box 3
King's Somborne
Stockbridge
SO20 6NE
UK
Tel: 44 1794 388158
Fax: 44 1794 388129

John Gordon International
Hillcrest House
4 Market Hill
Maldon
Essex CM9 4PZ
UK
Tel: 44 1621 858944
Fax: 44 1621 857733
E-mail: admin@johngordon.co.uk

Kemutec Group Ltd
Springwood Way
Macclesfield
Cheshire
SK10 2XA
UK
Tel: 44 1625 412000
Fax: 44 1625 412001

Mitchell Dryers Ltd
Denton Holme
Carlisle
Cumbria
CA2 5DU
UK
Tel: 44 1228 534433
Fax: 44 1228 401060

Natural Resources Institute
University of Greenwich
Chatham Maritime
Kent
ME4 4TB
UK
Tel: 44 1634 880088
Fax: 44 1634 883704

Pascall Engineering Company
51 Gatwick Road
Crawley
West Sussex
RH10 2RS
UK
Tel: 44 1293 525166
Fax: 44 1293 536214

Regis Machinery (Sales) Ltd
9B Arun Business Park
Bognor Regis
West Sussex
PO22 9SX
UK
Tel: 44 1243 825661
Fax: 44 1243 829364

Scotmec (Ayr) Ltd
1a Whitfield Drive
Heathfield
Ayr
KA8 9RX
UK
Tel: 44 1292 289999
Fax: 44 1292 610940

Silsoe Research Institute
Wrest Park
Silsoe
Bedford
MK45 4HS
UK
Tel: 44 1525 860000
Fax: 44 1525 860156

Status Instruments Ltd
Green Lane
Tewkesbury
Gloucestershire
GL20 8HE
UK
Tel: 44 1684 296818
Fax: 44 1684 293746
E-mail: sales@status.co.uk

Steele & Brodie Ltd
Beehive Works
Kilmany Road, Wormit
Newport-on-Tay
Fife DD6 8PG
UK
Tel: 44 1382 541728
Fax: 44 1382 543022
E-mail: steele&brodie@taynet.co.uk

Teacraft Ltd
PO Box 190
Kempston
Bedford
MK42 8DQ
UK
Tel: 44 1234 852121
Fax: 44 1234 853232
E-mail: teacraft@aol.com

Technical Sales International
PO Box 17
Lewes
East Sussex
BN7 3LS
UK
Tel: 44 1273 474366
Fax: 44 1273 477787

West Meters Ltd
Western Bank Industrial Estate
Wigton
Cumbria
CA7 9SJ
UK
Tel: 44 16973 44288
Fax: 44 16973 44616

Winkworth Machinery Ltd
Willow Tree Works, Swallowfield Street
Swallowfield
Nr Reading
Berkshire RG7 1QX
UK
Tel: 44 118 988 3551
Fax: 44 118 988 4031
E-mail: info@mixer.co.uk
Web: www.mixer.co.uk

North America

United States of America

C.S. Bell Co.
170 West Davis Street
PO Box 291
Tiffin
Ohio 44883
USA
Tel: 1 419 448 0791
Fax: 1 419 448 1203

The Essential Oil Company
1719 SE Umatilla St
Portland
Oregon 97202
USA
Tel: 1 800 729 5912
Web: www.essentialoil.com

Floragenics Distillation Systems
12840 Mt Echo Drive
Ione
CA 95640
USA
Tel: 1 877 446 3567
E-mail: info@floragenics.com
Sales@floragenics.com
Web: www.floragenics.com

The Grindmaster Corporation
4003 Collins Lane
Louisville
KY 40245
USA
Tel: 1 502 425 4776

International Ripening Company
1185 Pineridge Road
Norfolk
Virginia
23502–2095
USA
Tel: 1 757 855 3094
Fax: 1 757 855 4155
E-mail: info@QAsupplies.com
Web: www.qasupplies.com

Lehman Hardware & Appliances
One Lehman Circle
PO Box 41
Kidron
Ohio 44636
USA
Tel: 1 330 857 5757
Fax: 1 330 857 5785
E-mail: GetLehmans@aol.com

The Sausage Maker, Inc.
1500 Clinton St, Building 123
Buffalo
New York 14206
USA
Tel: 1 880 490 8525
Fax: 1 716 824 6465
E-mail: sausmaker@aol.com
Web: www.sausagemaker.com

Tom Industries
990 Aster Avenue
PO Box 800
El Cajon
California 92020
USA
Tel: 1 619 440 7779
Fax: 1 619 440 1810

South America

Argentina

Scheitler
Second hand equipment
Argentina
Web: www.scheitler.com.ar

Brazil

Brasholanda S/A – Equipamentos Industriais
Rua Brasholanda 01
Caixa Postal 1250
80001–970
Curitiba – PR
Brazil
Tel: 55 41 3662627
Fax: 55 41 2668234
E-mail: export@brasholanda.com.br
Web: www.brasholanda.com.br

ECIRTEC
Rua Padre Anchieta
No 7–61-Jardim Bela Vista
17060–400-BAURU-SP
Brazil

Colombia

Mecanicos Unidos
Carrera 42 No. 33–173 (Itagui)
Apartado Aereo 35–23
Medellin
Colombia
Tel: 574 372 2000/0147/1743
Fax: 574 281 4519/373 1747
E-mail: afmejia@epm.net.co

Peru

ANLIN
Av. Gran Chimu 648–650
Zarate
Lima
Peru
Tel: 51 14 375 0767
Fax: 51 14 375 0723

DISEG (Diseno Industrial y Servicios Generales)
Av. Jose Carlos Mariategui 1256
Villa Maria del Triunfo
Lima
Peru
Tel: 51 14 283 1417

FAINSA (Fabricantes En Acero Inoxidable SA)
480 Carmen de La Legue-Renoso
Callao
Peru
Tel: 51 14 452 4504/464 6180
Fax: 51 14 452 1552
Contact: Jiron Pacifico

FAMET IUGS SRL
Jr Yungay 1663
Chacra Rios Norte
Lima
Peru
Tel: 51 14 425 5696
Contact: Juan Lopez Ramirez

Industria Peruana Comercializadora Techno Pan Equipamiento Integral
Jr Las Limas 257
Urb Canto Bello
San Juan de Lurigancho
Peru
Tel: 51 14 375 0767
Fax: 51 14 375 0723

Industrias Technologicas Dinamicas SA
Av. Los Platinos 228
URB Industrial Infantas
Los Olivios
Lima
Peru
Tel: 51 14 528 9731
Fax: 51 14 528 1579

Servifabri SA
JR Alberto Aberd
No 400 Urb Miguel Grau (Ex Pinote)
San Martin de Porres
Lima
Peru
Tel/Fax: 51 14 481 1967

Vulcano Technologia Aplicada Eirl
Jiron Libertad 1970
El Tambo
Huancayo
Peru
Tel: 51 64 249 366

REFERENCE SECTION

Glossary

Absorption Uptake of moisture by dry foods.

Acid A liquid having a pH lower than 7. Acid is added to manufactured foods to improve flavour and to balance excessive sweetness. Acidic conditions (below pH 4.2) inhibit the growth of many food-spoilage and food-poisoning bacteria.

Acid food A food with a pH lower than 4.6.

Additives Chemicals, both synthetic and natural, that are added to foods to improve their eating quality or storage life. Additives that are permitted for use within the European Union are given E numbers.

Adulterants Chemicals that are intentionally added to food which are forbidden by law.

Alkali The opposite of an acid – a liquid having a pH above 7.

Amino acids Building blocks of proteins, containing an amino group ($-NH_2$) and a carboxyl group ($-COOH$). Amino acids are linked to form peptides, polypeptides and proteins. There are 22 different amino acids. Eight of these cannot be synthesized in the body and must be included in the diet. They are known as essential amino acids.

Amylases Enzymes that break down starch into smaller subunits (maltose, dextrin and glucose).

Antioxidants Substances that prevent rancidity developing in fats by either absorbing oxygen or preventing chemical changes that are involved in rancidity. Antioxidants which absorb oxygen include ascorbic acid (vitamin C), gallates and tocopherols (vitamin E). Chemical reactions involved in rancidity are prevented by BHA (butylated hydroxyanisole) or BHT (butylated hydroxytoluene).

Ascorbic acid Vitamin C. It occurs naturally in many fruits and vegetables. It is present in both oxidized and reduced forms. The latter is a powerful reducing agent and can be used as an antioxidant. Ascorbic acid is used in a number of products to protect flavours and fats against oxidation.

Benzoic acid A permitted preservative in a number of products. Like the benzoates, it is widely used in soft drinks and similar products, but use is declining because of allergic reactions in some people.

Blanching A short heat treatment carried out on vegetables prior to canning, freezing or drying. The main objectives are to inactivate enzymes and preserve colour. The efficiency of blanching can be tested using the peroxidase test. The main methods of blanching are water and steam blanching.

Browning reactions There are two main types of browning reaction.

Non-enzymic browning includes the following: Caramelization, the breakdown of sugars at high temperatures to produce brown products. Ascorbic acid oxidation, a slow process in some fruit juices that leads to the production of brown pigments. Maillard reaction, a reaction between reducing sugars and amino acids or proteins that results in the formation of brown melanoidin pigments. This reaction is quicker at high temperature and high pH.

Enzymic browning commonly occurs in fruit and vegetables that are bruised, cut and processed to expose surfaces to oxygen. It is the result of a reaction between enzymes (known as polyphenolases) and phenolic compounds that are present in the fruit.

Browning can be prevented by using sulphites or sulphur dioxide, by inactivating enzymes (for example by blanching) or by removing one or more of the reactants (for example, by removing sugar) to prevent the Maillard reaction.

Canning The process by which food is hermetically sealed in a container and then heat processed. The heat process is sufficient to kill all micro-organisms and their spores. A few spores may survive so the product is termed commercially sterile.

Caramel Sugar heated above its melting point. Caramel is a range of brown substances of variable composition that is used as a brown colour for foods and as a flavouring agent.

Caramelization The process by which caramel is produced.

Carbohydrates A food component, such as starch, found mostly in plant materials. Complex carbohydrates (such as starch) are built up from simple sugars (monosaccharides), including glucose, fructose and galactose, and disaccharides (sucrose, maltose).

Case hardening A problem occurring in the dehydration of some foods, e.g. fruits, when a hard skin is produced on the surface which then inhibits complete dehydration. It is usually caused by high temperatures, soluble solids and the denaturing of proteins on the surface of the produce.

Casein The main milk protein, representing about 3 per cent of milk. Casein is precipitated by acid at about pH 4.6 and is coagulated by rennet in cheese making.

Chilling A method of extending the storage life of a product by lowering its temperature to between –1 and 8 °C, but not by freezing the product.

Chlorine The component of bleach (or hypochlorite) that is used to sterilize water or equipment.

Clarification The process of removing suspended or colloidal material from liquid products such as wine. Centrifugation, filtration and enzymes can be used for the process.

Clean processing A term used to describe food processes that produce only minimal pollution and consume little energy or water.

Conching A process used in traditional chocolate manufacture to help develop the right texture.

Consumer The person or household that is the final buyer of a product.

Contaminants Materials that are accidentally included in a processed food.

Control point Stages in a process where a lack of control can affect the quality of a product.

Critical control point Stages in a process where a lack of control can affect the safety of a product.

Curd proteins Clotted proteins that are produced by the action of rennet on milk. Casein is modified by the rennet, allowing calcium to react with it and to coagulate. The liquid whey is drained off to leave the curd, which is transformed into cheese.

Curing A term usually applied to meat which involves the development of colour, flavour and enhanced keeping qualities. Curing brine (which contains salt, sodium nitrate, some sodium nitrite and sugar) is added to meat. The nitrate is converted to nitrite, which combines with pigments in the meat to produce the characteristic pink/red colour.

Customer A person, firm or institution that buys a product.

Demand The amount of goods that customers want or need to buy.

Denaturation A term that refers to the uncoiling of protein chains, caused by heat, changes in pH, agitation and sometimes light.

Detergents Chemicals that assist in removing soils from equipment or foods.

Dextrins Breakdown products of starch, which are soluble long chains of α-glucose units. They are formed when bread is toasted and are often used as edible adhesives.

Dextrose equivalent (DE) A value denoting the degree of conversion of starch into glucose. The higher the DE, the more glucose (dextrose) is present.

Disaccharides Sugars, such as sucrose, maltose and lactose, that are formed by the combination of two monosaccharides:
sucrose = fructose + glucose
lactose = glucose + galactose
maltose = glucose + glucose

Distribution channel The people or organizations that handle products as they are moved between a producer and a consumer.

Drying (dehydration) The removal of water from a food to preserve it.

Emulsifying agents Substances that enable the production of a stable dispersion of oil in water or vice versa. Examples include glycerol monostearate (GMS), lecithin, egg yolk and whey proteins. Some are more soluble in oil than water, but all orientate themselves at the interface between oil and water, and prevent droplets of oil from coalescing and separating. Emulsifiers are used in the production of mayonnaise and peanut butter, among other foods.

Enzymes These protein substances are naturally occurring catalysts found in the cells of plants and animals. If they are not inactivated they can produce undesirable changes in processed foods such as changes in colour, flavour and texture.

Extraction of flour This indicates the yield of flour obtained when milling wheat. Wholemeal flour is 100 per cent extraction (i.e. the whole wheat grain). White flour is about 72 per cent extraction, as the bran and germ are removed.

Extrusion This is a modern process of making snack foods and breakfast cereals, particularly corn- and potato-based products. The food material is heated under pressure in a barrel-shaped piece of equipment with either a single or double internal revolving screw. The revolving screw takes the product through the barrel to a die from which the product is released. The sudden pressure drop on release from the die makes the product expand rapidly to produce a light, porous texture. Various dies are used to give products of different shapes and sizes.

Fats Found in animals and plants, components of foods that provide energy.

Fermentation The metabolism of organic compounds in the absence of oxygen. Yeast ferments sugar to produce alcohol. Lactic acid bacteria ferment lactose in milk during yoghurt manufacture.

Filtration A process to remove particles from a liquid by passing through a cloth or other type of filter.

Free radicals Highly reactive chemical species formed from molecules with unpaired electrons. They are involved in initiating and perpetuating the reactions which cause fat to become rancid. The antioxidants BHA (butylated hydroxyanisole) and BHT (butylated hydroxytoluene) produce stable free radicals that stop rancidity reactions.

Freezing The conversion of water from the liquid state to solid ice. Several methods of freezing include the following:
Immersion – food placed in a very cold brine.
Plate – freezing by contact with a cooled flat surface.
Blast – cold air blown over the product to freeze it.
Cryogenic – liquid gases, nitrogen or carbon dioxide are sprayed over food.

Gelatinization The process by which a gel is formed. In the case of starch a large quantity of water is absorbed and the starch eventually forms a three-dimensional network. Starch from different sources gelatinizes at certain temperatures. Starches higher in amylose gelatinize at lower temperatures than those rich in amylopectin. The latter produces more stable gels.

Glucose A monosaccharide also known as dextrose. It is a reducing sugar with about 70–80 per cent the sweetness level of sucrose.

Glucose syrup A colourless syrup produced by the hydrolysis of starch. It contains glucose, maltose and dextrins and is used as a sweetening agent in confectionery.

Gluten The protein of wheat flour, made up of glutenins and gliadins. It is responsible for the extensibility and elasticity of dough in bread making. Strong flours are richer in gluten.

Glycerol monostearate (GMS) An emulsifier made from glycerol and one fatty acid (stearic acid). It is commonly used in the food industry as a general purpose emulsifier.

Grading The assessment of a number of attributes (size, weight, colour) to get an indication of overall quality of food. It is often confused with sorting, which is the separation of food into categories according to one or more physical attributes. Sorting can be used as part of a grading operation, but not vice versa. Grading is carried out by trained operators.

Gross weight The weight of a food item, including all packaging and/or non-edible parts (for example, peel and stones).

Hazard analysis The identification of ingredients, storage conditions, packaging and human factors which are potentially hazardous and may affect the safety of a food.

High-temperature short time (HTST) Describes heat processes such as pasteurization and sterilization of milk that use plate heat exchangers. The process allows the rapid heating of a product with minimal flavour changes but full bactericidal effect of the heat.

Homogenization A process applied to milk to break down fat into small stable droplets, which do not cream off. The normal method is to use a pressure homogenizer, which forces the milk through a small orifice under pressure.

Icing A method of chilling, used particularly for fish. Melting ice is better than solid ice as it absorbs latent heat and achieves a greater cooling effect.

Inversion The splitting of sucrose into its component monosaccharides – glucose and fructose.

Invert sugar A mixture of equal amounts of glucose and fructose. It is sweeter than sucrose and is added to confectionery for its sweetness. It occurs naturally in honey, jams and fruit juices.

Lactic acid A product of lactic acid bacteria (*Lactobacillus*) during fermentation, responsible for reducing the pH, for example during the fermentation of lactose in milk to give acidity and the flavour of sour milk. It is also produced during the preparation of pickles. Lactic acid is important in meat as glycogen is converted to lactic acid post mortem and is responsible for a pH of meat of around 5.4.

Lecithins Fatty substances belonging to the phospholipids. They are made of glycerol, fatty acids, phosphoric acid and choline. Lecithins are naturally occurring emulsifying agents found in egg yolk, milk, soya and peanuts, and are used in the preparation of mayonnaise and chocolate, among others. Commercially they are made from soya beans.

Listeria A group of bacteria that can give rise to a number of diseases, including meningitis.

Low acid food A food with a pH greater than 4.6.

Maillard reaction A browning reaction which is non-enzymic, resulting from the condensation of reducing sugars and amino acids or proteins. It occurs as the result of heating during cooking and processing, but also takes place slowly in some stored products, such as dried milk, if their moisture contents rise above 5 per cent. The reaction is accelerated at higher pH and retarded by the addition of acid or sulphur dioxide. The brown pigments formed are known as melanoidins.

Maltose A disaccharide and reducing sugar made from two glucose units. It is produced during the malting process (during brewing) from starch in barley grains.

Market A term derived from the physical marketplace, but now meaning any situation that brings together buyers and sellers of a product.

Marketing The series of activities to identify customers and then satisfy their needs by providing them with the products they want.

Market research Finding out about the types of people who buy particular products and why they buy them in order to identify market opportunities.

Market size The total amount of a product that is bought per month or per year in volume terms.

Market value The total amount of money spent on a product by consumers per month or per year.

Minerals Particular types of chemicals present in foods in trace amounts and essential for health.

Moisture content The amount of water in a food.

Monosaccharide Group name for the simplest sugars, ranging from three to seven carbon atoms. The most common ones are glucose, fructose and galactose.

Net content The weight of food in a container, excluding the weight of the package.

Nutritional labelling Information about the nutrients in a food displayed on the label.

Oils Fats that are liquid at room temperature. They usually contain a greater number of unsaturated fatty acids, which lowers the melting point.

Oxidation A reaction that involves the addition of oxygen to a compound, or the removal of electrons from an atom. It is always accompanied by a reducing reaction.

Pasteurization Heating food to below 100 °C to preserve it without substantial changes to its quality. Milk is pasteurized by heating to 72 °C for 15 seconds.

Pectin A chemical found in most fruits that can be used to form a gel (e.g. in jam making). Rapid-set pectin has a high methyl galacturonate content and sets rapidly, so is used to suspend fruit in jam.

Peroxidase test A test used for the efficiency of blanching. Peroxidase is a very heat-resistant enzyme so its denaturation by heat can be assumed to have included the denaturation of other enzymes present in the food.

pH A scale from 1 to 14 that is used to measure acidity (1–6), neutrality (7) or alkalinity (8–14).

Phenolic compound A complex compound that contains phenol groups (hydroxyl OH groups that are directly attached to a benzene ring). Phenolic compounds are common in plants in the form of tannins, lignins and flavanoids.

Pollution Unwanted changes to water, air, soil etc. that can cause harm to life.

Polysaccharides Long chains of monosaccharides joined together by glycosidic links. Starch is an example of a polysaccharide, made from chains of glucose units.

Preservatives Substances capable of preventing the spoilage of food by the action of micro-organisms. There is a permitted list for use in foods, the main two being sulphur dioxide and benzoic acid.

Preserves A range of products preserved by the osmotic effect of high sugar concentrations. The main examples are jams, marmalades, conserves and candied fruit.

Product development Creating or modifying a product to make something different for new or existing markets.

Proteins Chemicals in foods that are made from amino acids and are used by the body for growth and repair.

Quality assurance (QA) Management procedures that predict and prevent unwanted changes to the eating quality of foods, or prevent foods becoming unsafe.

Quality control (QC) Testing procedures that measure whether a food has the expected quality.

Rancidity A chemical change in fats and oils brought about by either oxidation or hydrolysis. It leads to the production of odours and flavours associated with the deterioration of fat.

Reducing reactions A reaction in which an electron is added to an atom or ion. Commonly this involves the removal of oxygen from one compound, which becomes reduced, while another compound is simultaneously oxidized.

Reducing sugar Sugars which contain a potential aldehyde group (-CHO). Some examples are glucose, maltose and lactose. Some sugars, such as fructose, contain a ketone group but are still reducing sugars. Reducing sugars react with amino acids and proteins in the Maillard browning reaction.

Rennet An extract from a calf's stomach. It contains the enzyme rennin and some pepsin and is responsible for coagulating milk during cheese making.

Sanitation Cleaning procedures.

Screening A method of size separation or sieving. It is used for dry cleaning of foods and sorting.

Smoking The use of smoke to aid preservation of and give flavour to meat products and fish. The preservative effect comes from phenols, acids and aldehydes in the smoke. Smoking also causes surface dehydration of the product. In hot-smoking the fish is cooked at the same time as being smoked.

Soils An overall term used for all types of contaminating materials on foods or equipment.

Solvent extraction The use of solvents, such as petroleum ether, to extract oils from vegetable sources, particularly seeds. The solvent is then distilled to leave a crude oil which needs refining.

Sorting The separation of foods into categories based on a measurable physical quality such as colour, weight, size or shape.

Stabilizers Substances, often complex polysaccharides, that have the ability to absorb considerable quantities of water. This property makes them good thickening agents, many being able to produce gels. Most can act as emulsifiers and prevent fat separation.

Starch A polysaccharide made from chains of glucose. Two forms exist – amylose, which is a straight chain in the form of a coil, and amylopectin, which is a highly branched form.

Starter cultures Special cultures of bacteria incubated under ideal conditions to be added to foods to start fermentation, particularly in wines, bread, yoghurt and cheese.

Sterilants Chemicals that kill micro-organisms.

Sterilization The achievement, usually by heating, of a complete absence of life (i.e. removal of all micro-organisms).

Sucrose A common sugar, either from beet or cane. It is a disaccharide made from glucose and fructose and is non-reducing. A non-reducing sugar does not contain any potential aldehyde groups (CHO) and therefore cannot participate in reducing reactions.

Sulphur dioxide A permitted preservative in many products. It is usually added as sulphite or metabisulphite. It is used in fruit products, sausages and wines.

Tannins Phenolic compounds with a complex structure. They are responsible for the astringency of red wines, tea, coffee and apples. They give body and fullness to the flavour of a product.

Ultra heat-treated (UHT) Describes high-temperature, short time sterilization, usually carried out in plate heat exchangers. UHT milk is usually called long-life milk.

Vitamins Particular types of chemicals present in foods in trace amounts and essential for health.

Waste Unwanted materials.

Water activity (a_w) A measure of the amount of water in a food that is available for microbial growth. Levels of moisture are compared with pure water, which has an a_w of 1.0. Preservation methods such as dehydration, concentration and the addition of salt and sugar rely on lowering the water activity.

Whey proteins Proteins from milk after the removal of casein.

Yeasts Fungi which are involved in the fermentation and spoilage of sweetened or salted products.

Yield Weight of food after processing compared to the weight before processing.

References and further reading

Agrodok (1991) *Beekeeping in the Tropics*, Agromisa, Wageningen, Netherlands.

Ash, R. (1983) *Cheesecraft*, Tabb House, Padstow, Cornwall, UK.

Axtell, B. (2002) *Drying Food for Profit: A Guide for Small Businesses*, ITDG Publishing, UK.

Axtell, B. and Bush, A. (1991) *Try Drying It*, Intermediate Technology Publications, London, UK.

Axtell, B. and Fairman, R.M. (1992) *Minor Oil Crops*, FAO Agricultural Services Bulletin 94, FAO Publications, Rome, Italy.

Bathie, G. (2000) *Baking for Profit: Starting a Small Bakery*, ITDG Publishing, London, UK.

Battcock, M. and Azam-Ali, S. (1999) *Fermented Fruit and Vegetables: A global perspective*, FAO Agricultural Services Bulletin 134, FAO Publications, Rome, Italy.

Biss, K. (1988) *Practical Cheesemaking*, Crowood Press, Marlborough, Wiltshire, UK.

Board, P.W. (1988) *Quality Control in Fruit and Vegetable Processing*, FAO Food and Nutrition Paper 39, FAO Publications, Rome, Italy.

Campbell, B. M. (1987) 'The Use of Wild Fruits in Zimbabwe', *Economic Botany* 41, pp. 378–385.

Caplen, R.H. (1982) *A Practical Approach to Quality Control*, Hutchinson Publishing Group, London, UK.

Caribbean Development Bank (1987) *Small-scale Processing of Ground Meat and Sausages*, Caribbean Development Bank, Barbados, WI.

Carruthers, I. and Rodriguez, M. (1992) *Tools for Agriculture: A Guide to Appropriate Equipment for Small-holder Farmers*, Intermediate Technology Publications, London, UK.

Clucas, I.J. and Ward, A.R. (1996) *Post Harvest Fisheries Development: A Guide to Handling, Preservation, Processing and Quality*, Chatham Maritime, Kent, ME4 4TB, UK.

Colquichagua, D. (1994) *Vino de Fruta*, ITDG, Lima, Peru.

Commonwealth Secretariat International Bee Research Association (1979), *Beekeeping in Rural Development*, Food Production and Rural Development Division, Commonwealth Secretariat, London, UK.

Conroy, C., Gordon, A. and Marter A. (1995) *Development and Dissemination of Agro-processing Technologies*, Natural Resources Institute, UK.

Dauthy, M.E. (1995) *Fruit and Vegetable Processing*, FAO Agricultural Services Bulletin 119, FAO Publications, Rome, Italy.

Dietz, M. (1999) 'The Potential of Small-scale Food Processing for Rural Economies', *The ACP-EU Courier*, 174, European Commission, Brussels, Belgium.

Dillon, M. and Griffith, C. (1995) *How to HACCP: An Illustrated Guide*, MD Associates, 34a Hainton Avenue, Grimsby, UK.

Dubach, J. (1989) *Traditional Cheesemaking*, Intermediate Technology Publications, London, UK.

Dubach, S. (1992) *A Lifetime Cheesemaking in Developing Countries*, Sepp Dubach, Ecuador.

FAO (1990) *The Technology of Traditional Milk Products in Developing Countries*, FAO Animal Production and Health Paper 85, FAO Publications, Rome, Italy.

Fellows, P. (1997a) *Guidelines for Small-scale Fruit and Vegetable Processors*, FAO Technical Bulletin 127, FAO Publications, Rome, Italy.

Fellows, P. (1997b) *Traditional Foods*, Intermediate Technology Publications, London, UK.

Fellows, P. (2000) *Food Processing Technology: Principles and Practice,* 2nd edn, Woodhead Publishers, Cambridge, UK.

Fellows, P. and Axtell, B. (2002) *Appropriate Food Packaging*, ITDG Publishing, London.

Fellows, P., Axtell, B. and Dillon, M. (1995) *Quality assurance for small rural food industries*, Technical Bulletin 117, FAO Publications, Rome, Italy.

Fellows, P., Franco, E. and Rios, W. (1996) *Starting a Small Food Processing Business*, Intermediate Technology Publications, London, UK.

Fellows, P., Hidellage, V. and Judge, E. (1998) *Making Safe Food* (Book and Posters), 2nd edn, ITDG, Rugby, UK.

Fleet, G.H. (1998) 'The Microbiology of Alcoholic Beverages', *Microbiology of Fermented Foods*, Wood, B.J.B. (ed.), Blackie Academic and Professional, London, UK.

Head, S.W., Swetman, A.A., Hammonds, T.W., Gordon, A., Southwell, K.H. and Harris, R.V. (1995) *Small-scale Vegetable Oil Extraction*, Natural Resources Institute, UK.

Hippisley Coxe, A. and Hippisley Coxe, A. (1994) *The Book of Sausages*, Victor Gollancz, London, UK.

Hobbs, B. and Roberts, D. (1987) *Food Poisoning and Food Hygiene*, Edward Arnold, London, UK.

Howard, J. (1979) *Safe Drinking Water*, Oxfam Technical Guide, Oxfam, Oxford, UK.

IFST (1991) *Food and Drink: Good Manufacturing Practice: A Guide to its Responsible Management*, Institute of Food Science and Technology, 5 Cambridge Court, 210 Shepherd's Bush Road, London W6 7NL, UK.

Ihekoronye, A.I. and Ngoddy, P. O. (1985) *Integrated Food Science and Technology for the Tropics*, Macmillan Publishers, London, UK.

ILO (1985a) *Small-scale Processing of Beef*, Technology Series Technical Memorandum 10, ILO, Geneva.

ILO (1985b) *Technologies for Rural Women: Ghana*, Technical Manual 3, Fish Smoking, ILO, Geneva.

Johnson, S. and Clucas, I. (1996) *Maintaining Fish Quality: An Illustrated Guide*, Natural Resources Institute, Chatham, UK.

Karki, T. (1986) 'Some Nepalese Fermented Foods and Beverages', *Traditional Foods: Some Products and Technologies*, Central Food Technological Research Institute (CFTRI), Mysore, India.

Kindervater, S. (ed.), (1987) *Doing a Feasibility Study: Training Activities for Starting or Reviewing a Small Business*, OEF International, 1815 H Street NW, 11th Floor, Washington, DC, 20006, USA.

Kordylas, J.M. (1990) *Processing and Preservation of Tropical and Subtropical Foods*, Macmillan Publishing, London, UK.

Kramer, A. and Twigg, B.A. (1962) *Fundamentals of Quality Control for the Food Industry*, AVI Publishing, Westport, Conn., USA.

Krell, R. (1996) *Value-added Products from Beekeeping*, FAO Agricultural Services Bulletin 124, FAO Publications, Rome, Italy.

Lal, G., Siddappa, G.S. and Tandon, G.L. (1986) *Preservation of Fruits and Vegetables*, Indian Council of Agricultural Research (ICAR), India.

Lee, C., Steinkraus, K.H. and Reilly, P.J.A. (1993) *Fish Fermentation Technology*, United Nations Press, United Nations University, Korea.

Mortimer, S. and Wallace, C. (1998), *HACCP: A Practical Approach*, Chapman and Hall, London, UK.

Norman, G.A and Corte, O.O. (1985) *Dried Salted Meats: Charque and Carne-de-Sol*, FAO Animal Production and Health Paper 51, FAO Publications, Rome, Italy.

Obi-Boatang, P. and Axtell, B.L. (1995) *Packaging*, Food Cycle Technology Source Books, ITDG Publishing, London.

Pickford, J., Barker, P., Coad, A., Dijkstra, T., Elson, B., Ince, M. and Shaw, R. (eds.) (1995) *Affordable Water Supply and Sanitation*, Intermediate Technology Publications, London, UK.

Potts K. and Machell, K. (1995) *The Manual Screw Press for Small-scale Oil Extraction*, Intermediate Technology Publications, London, UK.

Ranken, M.D., Kill, R.C. and Baker, C.G.J. (1997) *Food Industries Manual*, Blackie Academic and Professional, London, UK.

Rhodes, A. and Fletcher, D. (1966) *Principles of Industrial Microbiology*, Pergamon Press, UK.

Rothwell, J. (1985) *Ice Cream Making: A Practical Booklet*, College of Estate Management, Reading University, UK.

Shapton, D.A. and Shapton, N.F. (1993) *Safe Processing of Foods*, Butterworth-Heinemann, Oxford, UK.

Shapton, D.A. and Shapton, N.F. (1994) *Principles and Practices for the Safe Processing of Foods*, Woodhead Publications, Cambridge.

Sprenger, R.A. (1996) *The Food Hygiene Handbook*, Highfield Publications, Doncaster, UK.

Steinkraus, K.H. (1996) *Handbook of Indigenous Fermented Foods*, Marcel Decker, New York, USA.

UNIFEM (1993a) *Fish Processing*, Food Cycle Technology Source Books, Intermediate Technology Publications, London, UK.

UNIFEM (1993b) *Oil Extraction*, Food Cycle Technology Source Books, Intermediate Technology Publications, London, UK.

UNIFEM (1994) *Cereal Processing*, Food Cycle Technology Source Books, Intermediate Technology Publications, London, UK.

UNIFEM (1996) *Dairy Processing*, Food Cycle Technology Source Books, Intermediate Technology Publications, London, UK.

Walker, K. (1995) *Practical Food Smoking*, Neil Wilson Publishing Ltd, UK.

Water Research Centre (1989) *Disinfection of Rural and Small-community Water Supplies*, Water Research Centre, Bucks, UK.

White, E.C. (1993) *Super Formulas: Arts and Crafts: How to Make More Than 360 Useful Products That Contain Honey and Beeswax*, Valley Hills Press, Starkville, USA.

Wood, B.J.B. (1985) *Microbiology of Fermented foods,* Elsevier Applied Science Publishers, UK

World Bank (1989) *Sub-Saharan Africa*, World Bank, Washington, USA.

Young, R.H. and MacCormac, C.W. (1986) *Market Research for Food Products and Processes in Developing Countries*, IDRC, Ottawa, Canada.

Improvements to this book

The second edition of *Small-scale Food Processing* is only as good as the information we can collect from manufacturers. You can help us to improve the quality of this guide by doing any of the following:

1. Fill in the short questionnaire that follows and send it to us. This is most important as it will tell us what you want from the guide.
2. If you use the book to contact a manufacturer, please tell them that you found the address in *Small-scale Food Processing*. They will then be more likely to supply us with information next time around.
3. Send us details of any manufacturers of appropriate equipment that do not appear in this edition – name, address, telephone number and e-mail or web address (if available) and, if possible, equipment brochures.
4. Please write to us with any suggestions, criticisms or comments you have on the book.
5. Let us know if you find any errors in this book.

It is only through your help that this unique information resource can be improved to provide you, the user, with what you require.

Questionnaire

Please help us to improve future issues of this publication by completing this questionnaire and sending to: Small-scale Food Processing, ITDG, Schumacher Centre for Technology and Development, Bourton on Dunsmore, Rugby, CV23 9QZ, UK.

Name
Position/Profession
Address
E-mail

please tick appropriate boxes

How did you find out about this book?	Word of mouth	☐
	Saw it in a bookshop/library	☐
	Referred to in another book	☐
	Advertisement	☐
	Books by post catalogue	☐
	ITDG Publications trade list	☐
	Other (please specify)	______
Why did you buy this book?	To find manufacturers' details	☐
	To find information on small-scale food processing	☐
	To find out about specific equipment	☐
	Other (please specify)	______

Which chapters are of most interest to you?

Which chapters are of least interest to you?

What would you like to see included in the book?

Do you have access to the internet? YES/NO*

Would you use the internet to search for/buy equipment? YES/NO*

*Delete as appropppriate

Index